Materials Forming, Machining and Tribology

Series Editor

J. Paulo Davim, Department of Mechanical Engineering, University of Aveiro, Aveiro, Portugal

This series fosters information exchange and discussion on all aspects of materials forming, machining and tribology. This series focuses on materials forming and machining processes, namely, metal casting, rolling, forging, extrusion, drawing, sheet metal forming, microforming, hydroforming, thermoforming, incremental forming, joining, powder metallurgy and ceramics processing, shaping processes for plastics/composites, traditional machining (turning, drilling, miling, broaching, etc.), non-traditional machining (EDM, ECM, USM, LAM, etc.), grinding and others abrasive processes, hard part machining, high speed machining, high efficiency machining, micro and nanomachining, among others. The formability and machinability of all materials will be considered, including metals, polymers, ceramics, composites, biomaterials, nanomaterials, special materials, etc. The series covers the full range of tribological aspects such as surface integrity, friction and wear, lubrication and multiscale tribology including biomedical systems and manufacturing processes. It also covers modelling and optimization techniques applied in materials forming, machining and tribology. Contributions to this book series are welcome on all subjects of "green" materials forming, machining and tribology. To submit a proposal or request further information, please contact Dr. Mayra Castro, Publishing Editor Applied Sciences, via mayra.castro@springer.com or Professor J. Paulo Davim, Book Series Editor, via pdavim@ua.pt

More information about this series at https://link.springer.com/bookseries/11181

Majid Yaghoubi · Hamed Tavakoli

Mechanical Design of Machine Elements by Graphical Methods

Springer

Majid Yaghoubi
Rojan Sanat Alborz
Tehran, Iran

Hamed Tavakoli
Department of Engineering for Crop Production
Leibniz Institute for Agricultural Engineering and Bioeconomy e.V. (ATB)
Potsdam, Germany

ISSN 2195-0911 ISSN 2195-092X (electronic)
Materials Forming, Machining and Tribology
ISBN 978-3-031-04328-4 ISBN 978-3-031-04329-1 (eBook)
https://doi.org/10.1007/978-3-031-04329-1

This Springer imprint is published by the registered company Springer Nature Switzerland AG
The registered company address is: Gewerbestrasse 11, 6330 Cham, Switzerland

Preface

Mechanical design of machine components requires performing calculations using formulas, which is usually a sophisticated and time-consuming procedure. This book aims to provide students and engineers in mechanical engineering and related disciplines, practicing engineers, technicians, and manufacturers with an easy-to-use reference, which is based on using graphs instead of the formulas for designing common machine elements. It is assumed that the user has had basic knowledge in Mechanics, Strength of Materials, and Materials Properties.

The contents are based on the basic methods used in mechanical engineering design that have been modified by the authors, as well as several years of practical experiences of the authors in designing and manufacturing machine elements and mechanisms, in academia and industry.

The book is composed of 23 self-contained chapters, covering design of various machine elements such as joints, bolts, power screws, shafts, keys, splines, couplings, belt drives, chain drives, gear drives, rolling bearings, and the like. In general, the chapters are structured into four sections, namely, *Introduction, Design method(s), Equations*, and *References*. In some of the chapters, *Supplementary material* is also included.

The *Introduction* gives a brief overview about the element(s). The *Design method(s)* section introduces the graphical method(s). The procedure of applying the method(s) is explained through several examples. Since the aim is to use graphs instead of sophisticated formulas, we put all the related equations at the end of each chapter, for those readers who are interested. At the end, the references used in the chapter are listed in the *References* section.

A special characteristic of this book is proposing a simple, rapid, and novel method for a rough design of some of the elements based on the shaft size. We refer to this method as *M.Y method*. The method is very useful for the maintenance and repair workers. They can quickly find solutions for replacing parts by applying the method.

All the graphs have been created using the DESMOS software.

The International System of Units (SI) is used in this book.

In spite of efforts of the authors for writing and preparing the contents of the book in a very careful manner, errors are unavoidable. Therefore, any suggestions and feedbacks that can help us to improve the book are warmly welcomed.

Dr. Tavakoli wishes to dedicate his contributions in this work to his family and to the memory of his father.

At the end, this work is dedicated to those who spend their life for peace and human welfare.

Tehran, Iran Majid Yaghoubi
Potsdam, Germany Hamed Tavakoli

Acknowledgements

The authors would like to warmly thank the companies Autodesk Inc., DAIDO KOGYO CO. (DID), Parker corp., Renold plc., and SKF Group, and the publishers Springer, Carl Hanser Verlag, Wiley, Infra-Engineering, Academy for giving free permissions for reusing some content provided by them.

Special thanks are also due to the developers of the DESMOS software.

Contents

1 Torque ... 1
1.1 Introduction ... 1
1.2 Design Method ... 1
1.3 Equations ... 3

2 Hertzian Contact Stress ... 5
2.1 Introduction ... 5
2.2 Design Method ... 6
2.3 Equations ... 8
Reference ... 8

3 Press-Connect Joints ... 9
3.1 Introduction ... 9
3.2 Design Methods ... 10
3.2.1 Estimation of the Minimum Interference (Δ_{min}) ... 10
3.2.2 Estimation of the Maximum Interference (Δ_{max}) ... 12
3.2.3 Standard Interferencen ... 14
3.2.4 Heating Elements for the Assembly ... 16
3.2.5 M.Y Method (The Proposed Method in This Book) ... 17
3.3 Equations ... 19
References ... 19

4 Welded Joints ... 21
4.1 Introduction ... 21
4.2 Design Methods ... 22
4.2.1 Axial Loading ... 23
4.2.2 Bending Loading ... 25
4.2.3 Torsional Loading (Welded Shaft) ... 30
4.2.4 Combined Loading ... 32
4.3 Equations ... 33
References ... 33

5 Bolts ... 35
5.1 Introduction ... 35
5.2 Design Methods ... 36
5.2.1 Method I ... 36
5.2.2 Method II ... 41
5.2.3 Method III (M.Y Method) ... 48
5.3 Equations ... 49
References ... 51

6 Power Screws ... 53
6.1 Introduction ... 53
6.2 Design Methods ... 54
6.2.1 Method I ... 54
6.2.2 Method II (M.Y Method) ... 60
6.3 Equations ... 62
References ... 63

7 Shafts ... 65
7.1 Introduction ... 65
7.2 Design Methods ... 66
7.2.1 Rough Estimation (M.Y Method) ... 66
7.2.2 Precise Design (Checking) ... 70
7.3 Equations ... 75
References ... 76

8 Rectangular Keys ... 77
8.1 Introduction ... 77
8.2 Design Methods ... 78
8.2.1 Method I ... 78
8.2.2 Method II ... 80
8.2.3 Method III (M.Y Method) ... 82
8.3 Supplementary Material ... 84
8.4 Equations ... 84
References ... 85

9 Splines ... 87
9.1 Introduction ... 87
9.2 Design Method ... 88
9.3 Equations ... 92
References ... 93

10 Conical Joints ... 95
10.1 Introduction ... 95
10.2 Design Methods ... 96
10.2.1 Method I ... 96
10.2.2 Method II ... 98
10.3 Equations ... 101

References ... 101

11 Couplings ... 103
11.1 Introduction ... 103
11.2 Design Method ... 103

12 Flanged Couplings ... 107
12.1 Introduction ... 107
12.2 Design Methods ... 108
12.2.1 Method I ... 108
12.2.2 Method II (M.Y Method) ... 111
12.3 Equations ... 112

13 Universal Joints ... 113
13.1 Introduction ... 113
13.2 Design Methods ... 114
13.2.1 Relationship Between Torque and U-joint Width ... 114
13.2.2 Estimation of the Bearing Life ... 115
13.3 Equations ... 117
References ... 117

14 Pin Joints ... 119
14.1 Introduction ... 120
14.2 Design Methods ... 120
14.2.1 Shear Pin ... 120
14.2.2 Clevis Joint ... 125
14.2.3 Dowel Pin ... 131
14.3 Equations ... 138
References ... 139

15 Shafts and Their Associated Elements All Together ... 141
15.1 Introduction ... 141
15.2 Design Method (M.Y Method) ... 142
15.3 Equations ... 149
Reference ... 152

16 V-Belts ... 153
16.1 Introduction ... 153
16.2 Design Methods ... 154
16.2.1 Method I ... 154
16.2.2 Method II (M.Y Method) ... 161
16.3 Belt Tensioning ... 161
16.4 Equations ... 163
References ... 164

17 Timing Belts ... 165
17.1 Introduction ... 165
17.2 Design Methods ... 166
17.2.1 Initial Estimate ... 166
17.2.2 Checking Estimations ... 172
17.2.3 Belt Installation ... 175
17.3 Equations ... 176
References ... 177

18 Chain Drives ... 179
18.1 Introduction ... 180
18.2 Design Methods ... 180
18.2.1 Rough Design ... 181
18.2.2 Accurate Design ... 183
18.2.3 Sag Adjustment for Roller Chains ... 187
18.2.4 Design Based on the Tensile Strength of Chains ... 188
18.2.5 Lubrication Method ... 190
18.2.6 M.Y Method ... 191
18.2.7 Estimating the Minimum Center Distance ... 192
18.3 Supplementary Graphs and Equations ... 193
18.4 Equations ... 194
References ... 196

19 Spur and Helical Gear Drives ... 197
19.1 Introduction ... 198
19.2 Design Methods ... 199
19.2.1 Estimating the Width and Pitch Circle Diameter of Pinion ... 201
19.2.2 Estimating the Teeth Number of Pinion, and Module ... 206
19.2.3 Specifying the Gear Quality ... 208
19.2.4 Specifying the Helix Angle ... 209
19.2.5 Determining the Center Distance ... 211
19.2.6 M.Y Method ... 213
19.3 Supplementary Material ... 214
19.3.1 The Material Properties of Gears ... 214
19.3.2 The Module and Center Distance for Automobile Gearboxes ... 216
19.3.3 The Transmission Ratio in Multi-Stage Gearboxes ... 216
19.4 Equations ... 219
References ... 221

20 Bevel Gear Drives ... 223
20.1 Introduction ... 224
20.2 Design Methods ... 224
20.2.1 The Pinion Design ... 224
20.2.2 Specifying the Gear Quality ... 233
20.2.3 M.Y Method I ... 234
20.2.4 M.Y Method II ... 236
20.3 Equations ... 237
References ... 240

21 Wormgear Drives ... 241
21.1 Introduction ... 242
21.2 Design Methods ... 242
21.2.1 Method I ... 243
21.2.2 Method II (Suggested Method) ... 247
21.3 Equations ... 251
References ... 252

22 Rolling Contact Bearings ... 255
22.1 Introduction ... 256
22.2 Design Methods ... 256
22.2.1 Equivalent Load ... 256
22.2.2 Bearing Life ... 258
22.2.3 Allowable Shaft Speed for Rolling Bearings ... 262
22.2.4 Tolerances ... 264
22.3 Supplementary Material ... 264
22.4 Equations ... 271
References ... 271

23 Hydraulic Systems ... 273
23.1 Introduction ... 274
23.2 Design Methods ... 274
23.2.1 The Piston and Rod Diameters of the Hydraulic Cylinder ... 275
23.2.2 The Maximum Force of the Hydraulic Cylinder ... 276
23.2.3 The Hydraulic Pump Flow Rate and the Cylinder Speed ... 277
23.2.4 The Diameter of Hydraulic Pipes and the Required Power of Pumps ... 278
23.2.5 The Maximum Allowable Stroke of Hydraulic Cylinders ... 280
23.2.6 The Length of Hydraulic Cylinders ... 281
23.2.7 The Hydraulic Cylinder Thickness ... 282
23.3 Equations ... 284
Reference ... 285

About the Authors

Dr. Majid Yaghoubi is the first author of the book. He obtained his Ph.D. in the field of Mechanical Engineering of Agricultural Machinery from University of Tehran, Iran, in 2015. During his study, he was an outstanding and ingenious student; he achieved Iran's Elites Foundation award in 2010. His main expertise and interest are designing and fabrication of mechanical mechanisms and machines. He has more than 14 years of experience working as designer of dozens of projects, and manager of technical engineering departments, R&D departments, and production lines, in various industries such as automotive, injection molding, oil and gas, heavy machines, steel, mining, petrochemical, agricultural machinery, etc. He has been also active in publishing papers in peer-reviewed journals and books. In addition, he has patented several mechanisms nationally and internationally.

Dr. Hamed Tavakoli is the second author of the book. He obtained his Ph.D. in the field of Mechanical Engineering of Agricultural Machinery from University of Tehran, Iran, in 2014. During his study, he was an outstanding student in his field of study in Iran. He received Iran's Elites Foundation award in 2011. His experiences has been mainly in academia. From 2014 to 2019, he worked as an Assistant Professor at Department of Mechanical Engineering of Biosystems, Arak University, Iran. He has been working as a Researcher at Leibniz Institute for Agricultural Engineering and Bioeconomy e.V. (ATB), Potsdam, Germany, since February 2020. His main expertise and interest are instrumentation, design and development of sensor systems for plant and soil sensing in precision agriculture. He is author of several papers in high-ranked peer-reviewed journals of his field of research.

Chapter 1
Torque

Nomenclature

n	Rotational speed, rpm
P	Power, kW
T	Torque, N m

1.1 Introduction

Estimating torque is necessary during design of rotating elements such as gears, pulleys, shafts, etc. To calculate the torque we need to know power and rotational speed of the element. Power equals the product of torque and speed. In the ideal case, we can assume that the power is the same throughout the transmission system. However, in reality, there are small losses due to factors like friction in the bearings and gears.

1.2 Design Method

If we assume that the power is constant, wherever the speed changes, the torque changes too. For estimation of torque, Eq. 1.1 (see Sect. 1.3) can be used.

Instead of Eq. 1.1, the graph of Fig. 1.1 can be applied. The following example shows the use of the graphs.

Example 1.1 In Fig. 1.2, an electromotor produces a power of 100 kW at a speed of 2000 rpm. The power is transmitted through a universal joint to a gearbox. In the

M. Yaghoubi and H. Tavakoli, *Mechanical Design of Machine Elements by Graphical Methods*, Materials Forming, Machining and Tribology,
https://doi.org/10.1007/978-3-031-04329-1_1

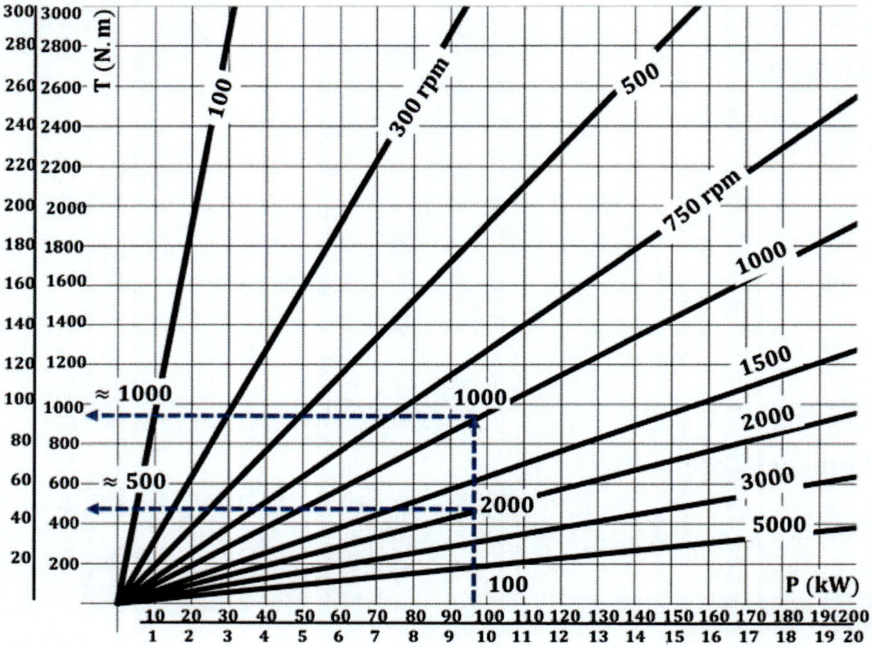

Fig. 1.1 A graph for estimating torque

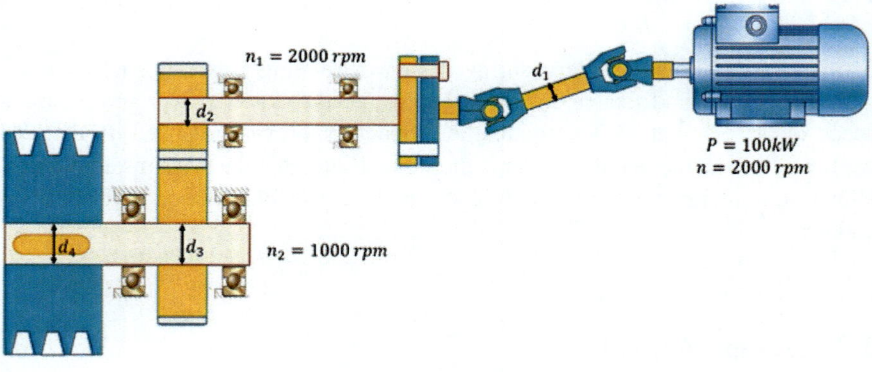

Fig. 1.2 Power transmission system of Example 1.1

gearbox, the speed is reduced to 1000 rpm and is transferred to the pulley. Find the torque on each part.

Solution

1. First, we separate all the components that have the same speed.

Group A: the electromotor, u-joint, input shaft of the gearbox and its gear, $n_1 = 2000$ rpm

Group B: the output shaft of the gearbox, its gear, and pulley, $n_2 = 1000$ rpm

2. Now, in Fig. 1.2, we select the power of 100 kW on the horizontal axis and go up vertically to intersect curves of 2000 and 1000 rpm. From the intersection points, we go to the left to get the torque that is applied to each component.

According to the diagram, the torque for the elements of group A is 500 N m and for group B is 1000 N m.

Since a greater torque is applied onto the output shaft and the components attached to it, they should be designed stronger.

1.3 Equations

$$T = 9550\frac{P}{n} \tag{1.1}$$

Chapter 2
Hertzian Contact Stress

Nomenclature

D	Diameter of the larger cylinder in contact, mm
d	Diameter of the smaller cylinder in contact, mm
F	Applied force, N
l	Contact length, mm
σ	Hertzian stress, MPa

2.1 Introduction

When two elements with curved surfaces make contact with each other and are pressed together, contact stresses (also called Hertzian contact stresses) are developed within the contact areas. Practical examples include the contact between wheels and rails, valve cams and tappets, mating gears and rolling bearings. Theoretically, the contact is either a point (when at least one of the elements is a sphere) or a line (when at least one of the elements is a cylinder).

Figure 2.1 shows different types of such contacts in which the Hertzian contact stresses can occur. High contact stresses may lead to failures such as cracks, pits and flaking in the surface material.

In this chapter, we only consider the line contact as it is more common and present graphs to simplfy designing of it.

M. Yaghoubi and H. Tavakoli, *Mechanical Design of Machine Elements by Graphical Methods*, Materials Forming, Machining and Tribology,
https://doi.org/10.1007/978-3-031-04329-1_2

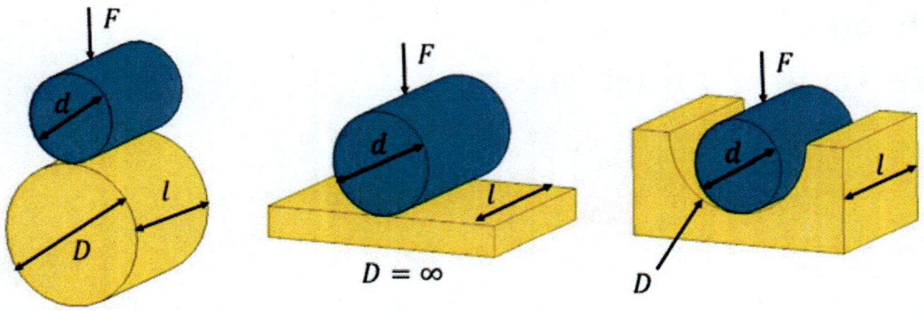

Cylinder on cylinder Cylinder on flat plate Cylinder in inner cylinder

Fig. 2.1 Different types of contacts in which the Hertzian stress occurs

2.2 Design Method

The Hertzian stress depends on the material and geometries of the contacting elements as well as the applied loads. To calculate the stress, the Hertz formula can be used (Jiang, 2019). However, for two elements which are made of steel, Eq. 2.1 (see Sect. 2.3) can be simply used for calculating the stress.

In case that one of the elements is flat, $D = \infty$ and therefore, $\frac{1}{n} = 0$. "+" is applied when the outer curved surface of the larger cylinder is in contact and "−" is used when the inner curved surface of the larger cylinder is in contact (Fig. 2.1).

Instead of Eq. 2.1 the graphs of Fig. 2.2 may be used. The following example shows the use of the graphs.

Example 2.1 Consider a crane that moves on a rail via rollers. Assume that on one of the rollers, a force of 2400 kg (≈24,000 N) is applied (Fig. 2.3). The rollers and the rail are made of ck45 steel with a hardness of 45 HRC. Check if the roller and rail can tolerate the load or not? The diameter and width of the roller are 50 and 60 mm, respectively and the width of the rail is 50 mm.

Solution

To solve this problem we follow the steps below:

1. In the second quadrant (the upper left corner) of Fig. 2.2, we find the force, 2400 kg (or 24,000 N). We go up from that point to cross the curve related to the roller with diameter of 50 mm.
2. Then we go right to cross the curve pertaining to width of 50 mm (we select the minimum width of the contacting elements).
3. Then we go down to reach the curve of cylinder on a flat surface (or rail) (). From that point, we go left to cross the vertical axis and find the stress (Hertzian stress) of the roller or rail that is 850 MPa.
4. In the fourth quadrant (the lower right corner) of Fig. 2.2, we find 45 HRC on the horizontal axis, and go up to cross the curve (For safety, we choose the lower

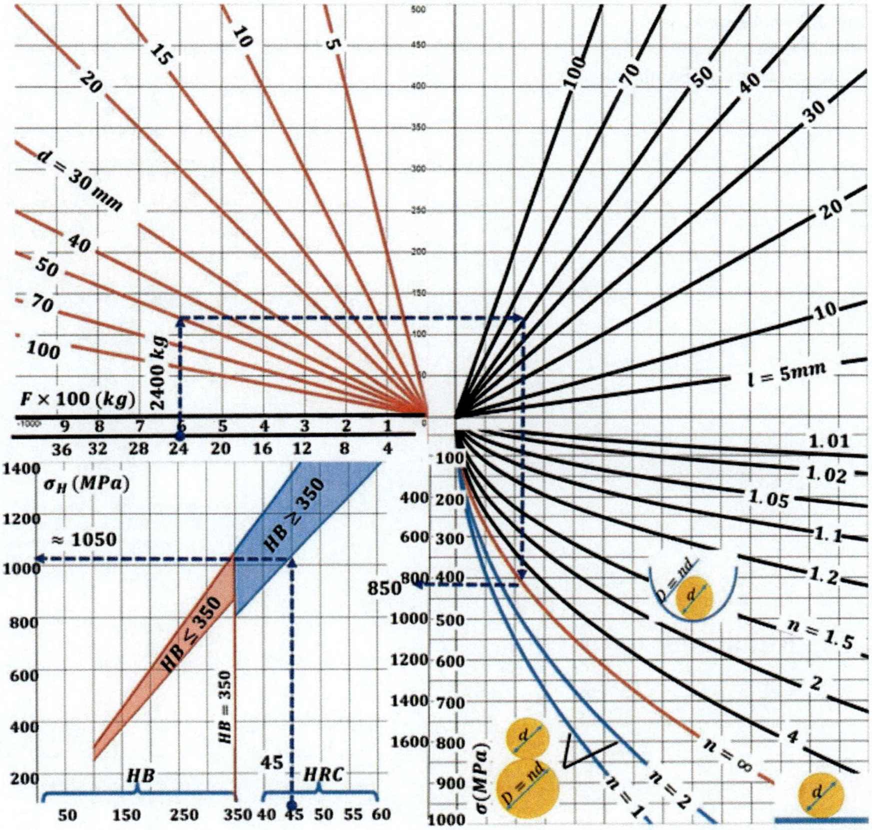

Fig. 2.2 Graphs for estimating Hertzian stress

limit of the graph). Then we go left to reach the vertical axis. The maximum allowable stress for the roller and rail is 1050 MPa.

Conclusion: as the estimated Hertzian stress (850 MPa) is lower than the maximum allowable stress for the elements (1050 MPa), they can tolerate the load.

However, to ensure a safe design, we apply a safety factor of 1.5. Therefore, the minimum values for the diameter of the roller and width of the rail should be:

$$d = 1.5 \times 50 = 75 \text{ mm}$$

$$l = 1.5 \times 50 = 75 \text{ mm}$$

Note: The number of cycles to failure at maximum allowable stress in Fig. 2.2 is 180×10^6 for H $\leq$ 350 HB and 3000×10^6 for H $\geq$ 350HB. For instance, in Example 2.1, after 3000×10^6 cycles the roller starts to fail.

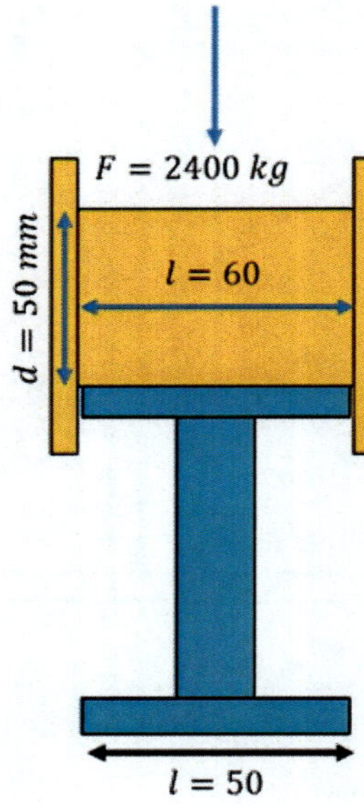

Fig. 2.3 Schematic of contact between a roller of a crane and a rail for Example 2.1

2.3 Equations

$$\sigma = 275^2\sqrt{\frac{F}{ld}}\left(1 \pm \frac{1}{n}\right) \tag{2.1}$$

Reference

Jiang W (2019) Analysis and design of machine elements, 1st edn. Wiley, New Jersey

Chapter 3
Press-Connect Joints

Nomenclature

D	Outer diameter of hub, mm
d	Outer diameter of shaft, mm
d_i	Inner diameter of shaft, mm
E	Elastic modulus, MPa
F_{axial}	Axial force, N
k	Unit pressure on the surface of joint, MPa
l	Connection length, mm
R_e	Yield strength, MPa
T	Torque, N m
Δ	Interference, μm
μ	Coefficient of friction
σ	Stress on surface of components, MPa

3.1 Introduction

Joining elements through press-connect creates durable and permanet connections that can transmit large and abruptly applied loads or fluctuating forces. Since this kind of joining does not require any connecting parts such as keys and consequently any keyways or grooves in which the stress concentration occours, it has a high design strength (operational strength) (Decker and Kabus 2014).

In this chapter, we present graphs that simplfy the press-joint design.

M. Yaghoubi and H. Tavakoli, *Mechanical Design of Machine Elements by Graphical Methods*, Materials Forming, Machining and Tribology,
https://doi.org/10.1007/978-3-031-04329-1_3

3.2 Design Methods

To design press-connect joints, Eqs. 3.1, 3.2, 3.3, 3.4 and 3.5 (see Sect. 3.3) can be used (also see Fig. 3.1a).

Instead of using the equations above, we propose a graphical method for designing press-fit joints a presented in the following subsections. In the graphical method, the coefficient of friction and connection length are assumed to be $\mu = 0.1$ and $l = d$, respectively.

3.2.1 *Estimation of the Minimum Interference* ($\Delta_{\min}$)

Example 3.1 shows use of the graphical method for rough calculation of the minimum interference.

Example 3.1 The cross-section view of a hollow shaft, which is connected to a hub by press-fit and is supposed to transmit a torque of 40 kg m is shown in Fig. 3.1b. Find the minimum interference for the press-fit joint.

Solution

To solve this problem we follow the steps below:

1 First we find the ratio of inner diameter to outer diameter (a) for both the shaft and hub:

$$\begin{cases} a_1 = \frac{d_i}{d} = \frac{30}{60} = 0.5 \\ a_2 = \frac{d}{D} = \frac{60}{150} = 0.4 \end{cases}$$

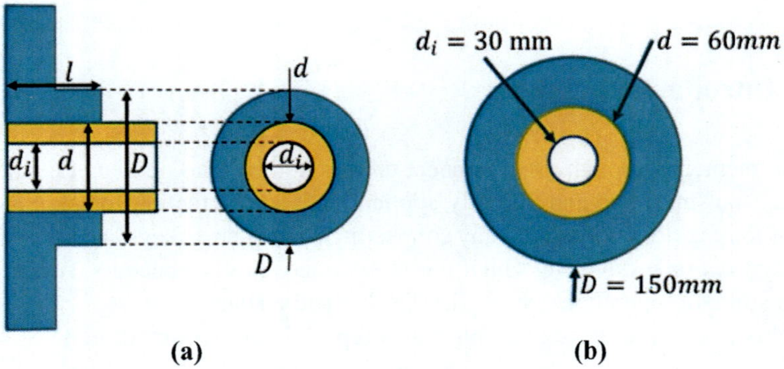

Fig. 3.1 **a**: Cross-section view of a hollow shaft connected to a hub by press-connect. **b**: The press-connect joint of Example 3.1

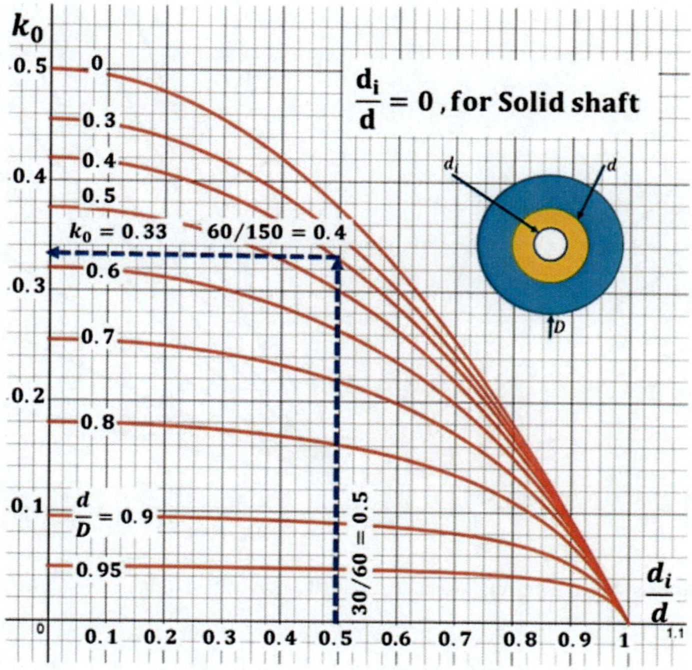

Fig. 3.2 A graph for estimating the parameter k_O

Then, we find the value of k_O from Fig. 3.2. On the horizontal axis, we locate $\frac{d_i}{d} = 0.5$ and go up to cross the curve of $\frac{d}{D} = 0.4$. We select the higher value ($a = 0.5$) as reference. Because the higher the ratio the weaker the element. We go left to obtain the corresponding k_O, that is:

$$k_O = 0.33$$

2 On the left horizontal axis of Fig. 3.3, we select $T = 40$ kg m and go up to reach the curve related to $d =$ *60* mm. From the point of intersection, we go right to cross the curve associated with $k_O = 0.33$. Then, going down to reach the right horizontal axis gives us the *minimum* interference fit:

$$\Delta_{min} = 10\,\mu\text{m}$$

Therefore, the shaft in Fig. 3.1 with the connection length of $l = d = 60$ mm can transmit the torque of 40 kg m, if the minimum interference is 10 μm. However, it is preferable to use the highest torque capacity of the joint, based on the strength of their material (see Sect. 3.2.2).

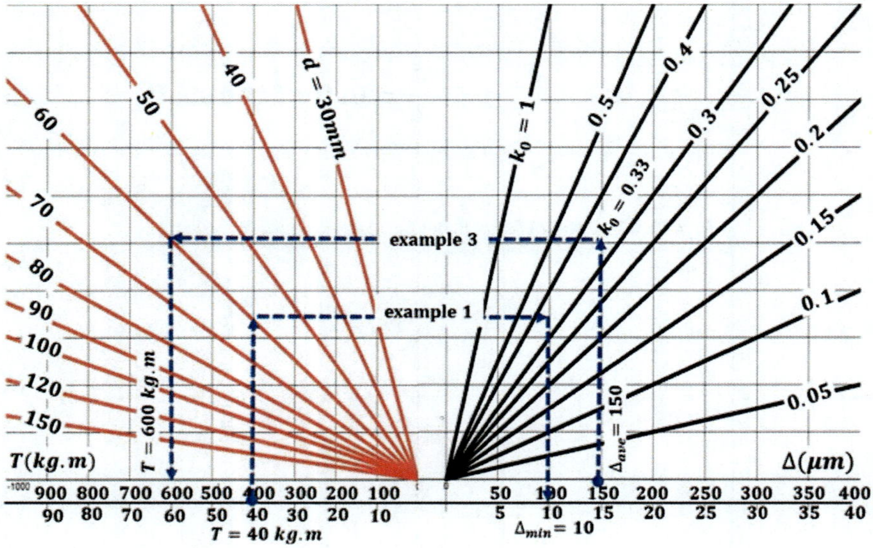

Fig. 3.3 Graphs showing the relationship between the transmitted torque and interference

3.2.2 *Estimation of the Maximum Interference (Δ_{max})*

In designing press-fit joints, to prevent damage in the elements, we should take into account that the interference should not exceed a specific amount. Therefore, we ought to obtain the maximum interference so that during the assembly, the stresses occurring at the shaft and hub do not exceed their allowable stresses.

We show how the maximum interference can be estimated by the proposed graphical method in Example 3.2.

Example 3.2 If the shaft and hub of Example 3.1 are made of 34CrMo4 steel with yield stress of $R_e = 550$ MPa (Table 3.1), estimate the maximum interference so that damages in the element do not occur.

Solution

1 First we calculate a for the shaft and hub:

$$\begin{cases} a_1 = \frac{d_i}{d} = \frac{30}{60} = 0.5 \\ a_2 = \frac{d}{D} = \frac{60}{150} = 0.4 \end{cases}$$

We select the higher value ($a = 0.5$) as reference. Because the higher the ratio the weaker the element.

2 In Fig. 3.4, from the lower vertical axis, we select $R_e = 550$ MPa and go right to cross the curve of $a = 0.5$.

Table 3.1 Yield strength of some material (Decker and Kabus; Maschinenelemente, 2014 Carl Hanser Verlag München)

			Heat treated								Case hardened
Material	St37	E295	C35	E335	Ck45	34CrMo4	25crMo4	34CrMo4	42CrMo4	50CrMo4	20MnCr5
R_e (MPa)	215	255	270	315	305	550	450	550	650	700	550

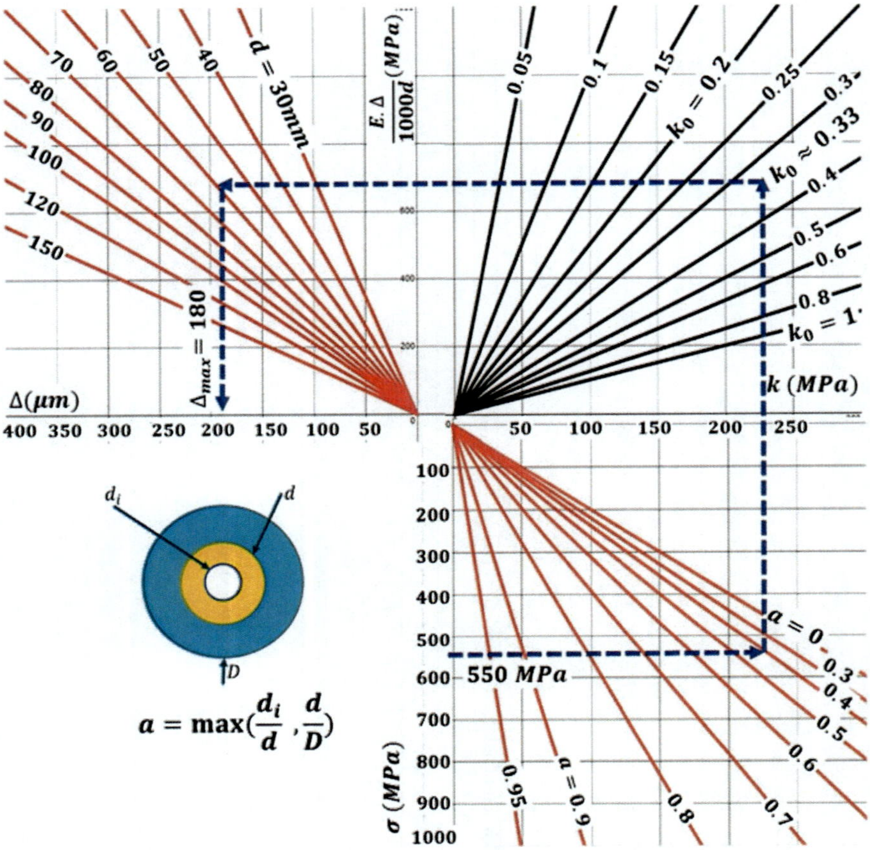

Fig. 3.4 Graphs representing the relationship between stress and interference

3 Then we go up to reach the curve of $k_O = 0.33$ that we obtained in Example 3.1.
4 From that point we go left to cross the curve of $d = 60$ mm and then down to obtain:

$$\Delta_{max} = 180\,\mu\text{m}$$

Thus, the value of the clearance should not exceed 180 μm.

3.2.3 *Standard Interferencen*

The interference is usually selected from standard values. For this purpose, Fig. 3.5 can be used. The interference should be selected so that its limits (minimum and maximum values) are between calculated (or estimated) ones. The following example illustrates the procedure.

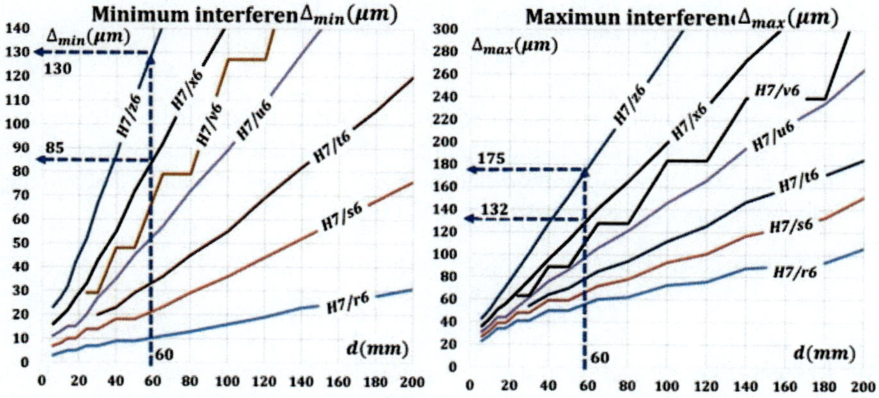

Fig. 3.5 Graphs for selecting standard interference

Example 3.3 Specify the standard interference for the press-fit joint of Examples 3.1 and 3.2 in which $\Delta_{min} = 10\ \mu m$ and $\Delta_{max} = 180\ \mu m$.

Solution

We should choose a tolerance that the minimum and maximum values lie between 10 and 180 μm. The closer to the maximum allowable value, ($\Delta_{max} = 180\ \mu m$), the better.

1 From Fig. 3.5, we select the *H7/* × *6* tolerance arbitrarily. In both sections (right and left), we choose the diameter of $d = 60$ mm, on the horizontal axis and go up to intersect the curve related to *H7/* × *6*; then, go to the left to achieve the minimum and maximum interference. The following values are obtained:

$$\Delta_{min} = 85\ \mu m$$
$$\Delta_{max} = 132\ \mu m$$

Since the values are between the minimum and maximum interference values estimated in Examples 3.1 and 3.2, respectively, the selected standard is an appropriate choice.

2 If we select *H7/z6*, that the maximum and minimum values can be:

$$\Delta_{min} = 130\ \mu m$$
$$\Delta_{max} = 175\ \mu m$$

As these values lie between the estimated Δ_{min} and Δ_{max}, this standard (*H7/z6*) is also a proper choice.

The average of Δ_{min} and Δ_{max} for *H7/z6* is about:

$$\Delta_{ave} = \frac{\Delta_{max} + \Delta_{min}}{2} = \frac{175 + 130}{2} \approx 150\,\mu\text{m}$$

The maximum interference of *H7/z6* ($\Delta_{max} = 175$ μm) is closer to the estimated one ($\Delta_{max} = 180$ μm). On the other hand, due to the Rz roughness error in the manufacturing process, we think that the *H7/z6* type is more suitable and reliable. Therefore, our choice is *H7/z6* type.

Now, we find the average torque-transmitting capacity of our design using $\Delta_{ave} = 150$ μm.

3 Refer to Fig. 3.3 and choose $\Delta_{ave} = 150$ μm, go up to intersect $k_0 = 0.33$, then go left to intersect $d = 60$ mm and go down. The corresponding average torque-transmitting capacity is:

$$T = 600\,\text{kg m}$$

Thus according to Examples 3.1, 3.2 and 3.3, the final results are: $d = 60$ mm, $l \geq d = 60$ mm, the interference: *H7/z6*

3.2.4 Heating Elements for the Assembly

For connecting the shaft and hub, instead of applying axial forces, making difference between their temperatures is often used. Mostly the hub is warmed up or the shaft is cooled down. To specify the temperature difference (Δt), Fig. 3.6 can be applied, as shown in Example 3.4.

Example 3.4 consider the press-fit joint of Example 3.3 in which the tolerance is *H7/z6* with $\Delta_{max} = 175$ μm. Determine the temperature of hub so that the assembly can be made without any force. Temperature of the shaft is 20 °C.

Solution

1 On horizontal axis of Fig. 3.6, we find $\Delta_{max} = 175$ μm, go up to reach the curve of $d = 60$ mm, and then go left to cross the vertical axis. The corresponding temperature difference is:

$\Delta t = 260$ °C

2 Since the temperature of the shaft is 20 °C, the temperature of the hub should be 280 °C (260 + 20 = 280 °C) so that they connect without any axial force.

Note: temperature of the element should not exceed 300 °C

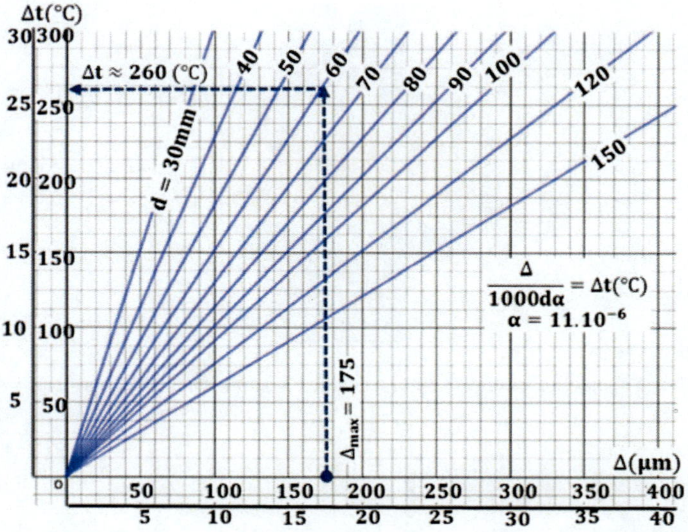

Fig. 3.6 A graph for determining the temperature difference for assembly the shaft and hub

3.2.5 M.Y Method (The Proposed Method in This Book)

In our proposed rough and simplified method, you are able to estimate diameter of the shaft and the interference simultaneously, based on the arrangement of the shaft. The assumptions for the method are:

1 The shaft and hub should be made of either ck45 steel or a stronger material.
2 The connection length should be greater than or equal to the shaft diameter ($l \geq d$). In addition, the hub outer diameter should be greater than or equal to two times the shaft diameter ($D \geq 2d$).
3 Solid shafts should be selected.

We explain the method in Example 3.5.

Example 3.5 The torque of 2000 N m is supposed to be transmitted from a gearbox to a conveyor. The output shaft of the gearbox is connected to the conveyor's shaft by flange coupling (pure torsion) (Fig. 3.7). If we connect the coupling to the shafts by press fit, specify the diameter of the coupling and the interference.

Solution

The proposed method is based on first obtaining the shaft diameter, then finding the interference so that the strength of the connection is higher than that of the shaft.

1 On the left horizontal axis of Fig. 3.8, we select $T = 2000$ N m, go up to reach the curve of "*pure torsion*" (because we assume that there is no misallignment between the two shafts), then go right to find the shaft's diameter. It is $d =$ 65 mm.

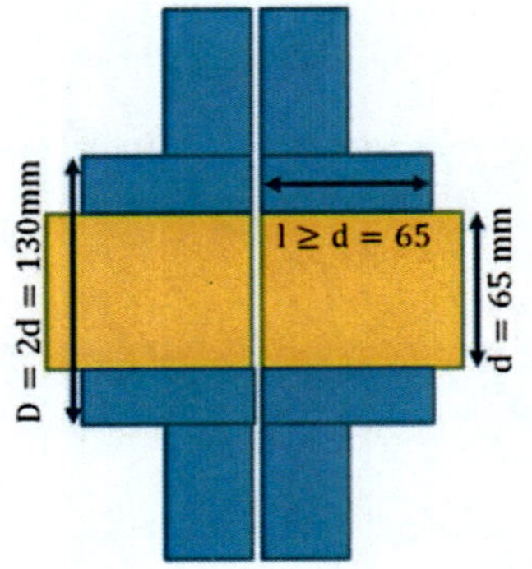

Fig. 3.7 Flange coupling for connection of a gearbox's output shaft and a conveyor's shaft

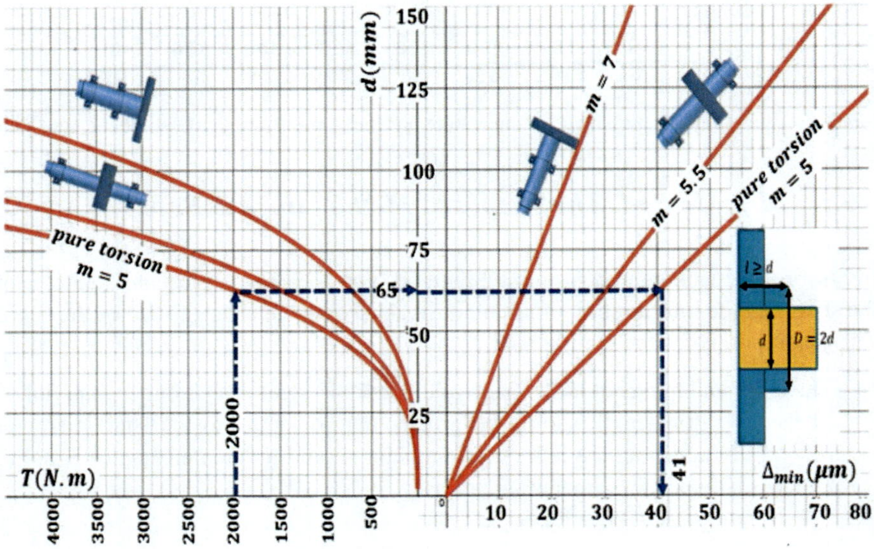

Fig. 3.8 Graphs for rough design of press-connect joints (M.Y method)

2 From the point of $d = 65$ mm, we continue going right to cross the curve of "*pure torsion*" on the right section of Fig. 3.8, then going down to the right horizontal axis gives us the minimum interference that is:

$$\Delta_{min} = 41\ \mu\text{m}$$

3 The maximum interference fit is:

$$\Delta_{max} = 0.001428d \xrightarrow{d=65\text{ mm}} \Delta_{\max} = 0.009282\text{ mm} = 92.82\ \mu\text{m}$$

4 Therefore, the dimensions of the coupling are as follows:

$$d = 65 \text{ mm}$$
$$D = 2 \times d = 130 \text{ m}$$
$$l \geq d \rightarrow l \geq 65 \text{ mm}$$
$$\Delta_{min} = 41\ \mu\text{m}$$
$$\Delta_{max} = 92.82\ \mu\text{m}$$

Note: If you use cast iron or aluminum for the hub instead of steel, the contact length should be $l \geq 2d$ and $l \geq 1.5d$ respectively

3.3 Equations

$$\mathrm{T} = \begin{cases} 0.001\ \mathrm{kl}\pi d\mu\frac{d}{2} \\ \text{if } \mu \approx 0.1 \text{ then } l = d \end{cases} = \frac{\pi \mathrm{kd}^3}{20000} \tag{3.1}$$

$$\sigma = \frac{2k}{1-a^2} \leq R_e, a = \left(\frac{d_i}{d} \text{ or } \frac{d}{D}\right) \tag{3.2}$$

$$F_{axial} = kl\pi \mathrm{d}\mu = \frac{2000}{\mathrm{d}}\mathrm{T} \tag{3.3}$$

$$k = \frac{E\Delta}{1000d} \times k_0 \rightarrow^{(for\ steel\ E=210000\ MPa)} = \frac{210\Delta}{d}k_0 \tag{3.4}$$

$$k_0 = \frac{1}{c_1 + c_2}, c_1 = \frac{1 + (d_i/d)^2}{1 - (d_i/d)^2}, c_2 = \frac{1 + (d/D)^2}{1 - (d/D)^2} \tag{3.5}$$

References

Decker K-H, Kabus K (2014) Decker Maschinenelemente Funktion, Gestaltung und Berechnung. Carl Hanser Verlag, München (in German)

Dunaev PF, Lelikov OP (2008) Designing of units and machine parts. Academy, Moscow (in Russian)

Hall A (1961) Theory and problems of machine design. McGraw-Hill, New York

Mott RL, Vavrek EM, Wang J (2018) Machine elements in mechanical design, 6th edn. Pearson Education, New York

Orlov P (1988) Fundamentals of Machine Design. Mechanical Engineering, Moscow (in Russian)

Wittel H, Muhs D, Jannasch D, Voßiek J (2013) Roloff/Matek Maschinenelemente Normung, Berechnung, Gestaltung. Springer Vieweg, Wiesbaden (in German)

Chapter 4
Welded Joints

Nomenclature

F	Axial force, N
J_w	Polar moment of inertia, mm^3
L	Distance between force and gravity center of weld, mm
l	The length of weld, mm
r	Distance between center of mass and farthest point of weld, mm
S	Linear stress, MPa
t	Thickness of welded elements, mm
z	The width of weld, mm
Z_w	Section module of weld, mm^2

4.1 Introduction

Welded joints are one of the most common types of permanent joints. Various welding processes are used for creating the joints. Fusion welding that fuses the base metal to form welds is one of the most widely used ones. Main welded joint designs are butt, lap, T-, edge and corner joints. Welds are named for the geometry of the edges of the parts to be joined. Table 4.1 shows different types of welded joints and welds. Basically, there are two types of weld seams; butt welds, which are used to connect elements within a plane and fillet welds, which are used to join components in different planes (Jiang 2019; Mott et al. 2018).

The calculations of welded joints are typically very time-consuming. However, in this chapter we provide diagrams that facilitate the calculations. Here, we only focus on fillet welded joints.

M. Yaghoubi and H. Tavakoli, *Mechanical Design of Machine Elements by Graphical Methods*, Materials Forming, Machining and Tribology,
https://doi.org/10.1007/978-3-031-04329-1_4

Table 4.1 Types of welded joints and weldsn (modified from Jiang (2019))

Welded joints	Welds
 Butt joint	 Single-V groove weld
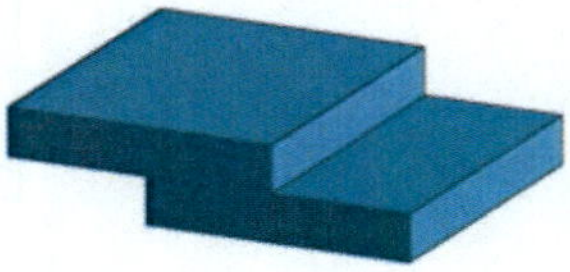 Lap joint	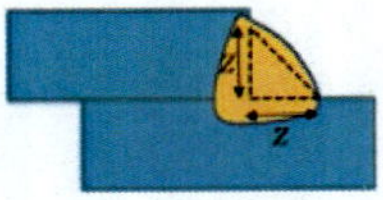 Single fillet weld
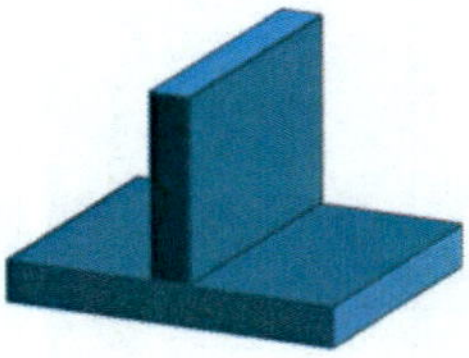 T-joint	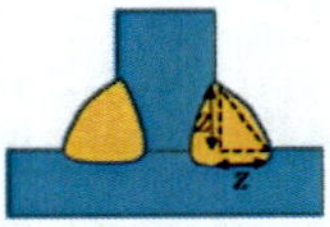 Double fillet weld
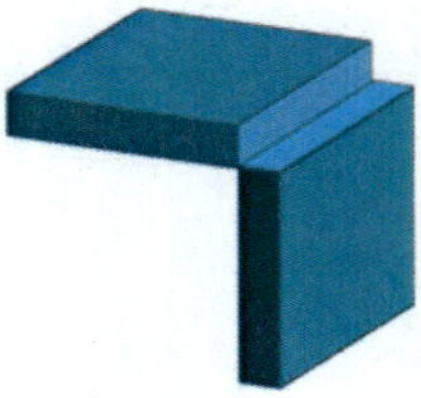 Corner joint	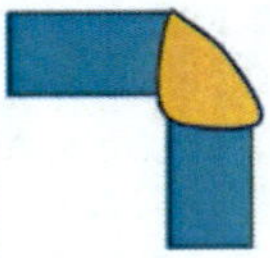 Corner weld
 Edge joint	 Edge weld

4.2 Design Methods

In welding, design usually includes determining the width of weld (z) so that the weld does not get ruptured. Equation 4.1 is used for this purpose (see Sect. 4.3).

It sometimes takes a long, confusing and time-consuming process to calculate the width of weld. However, using the proposed graphical method of this book, which is illustrated in the following subsections and examples, you can easily estimate it.

We consider designing weld for cantilevered bars (or plates) under four different loadings of axial (tension), bending, torsional and combined.

4.2.1 Axial Loading

For cantilevered bars (or plates) under axial loading, Eq. 4.2 (see Sect. 4.3) can be applied to calculate the linear stress of weld (S).

Example 4.1 A plate with a width and thickness of 300 and 30 mm, respectively, is supposed to be welded and be under load of 20,000 kg, on a crane (Fig. 4.1a). Find the width of weld (z) so that the weld can tolerate the load. How about if a round bar of diameter $d = 100$ mm, is used instead of the plate (Fig. 4.1b)?

Solution

1 First, we should determine the length of welds. In case of the plate (Fig. 4.1a), the length (l) is:

$$l = 200 + 300 + 200 = 700 \text{ mm}$$

And for the bar (Fig. 4.1b), as it is welded double (over and under the blue part), the length is:

$$l = 2 \times (\pi \times \text{d}) = 2 \times (\pi \times 100) = 628 \text{ mm}$$

2 Now, we calculate the linear stress. For converting the force to Newton (unit), we multiply it by 10:

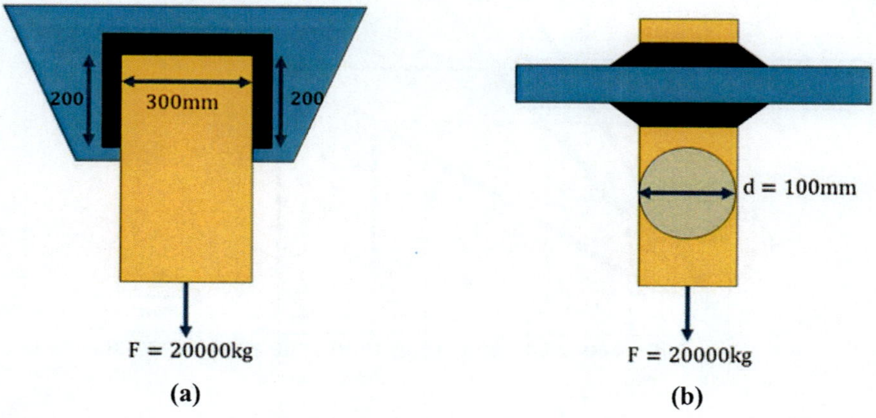

Fig. 4.1 A plate and a bar loaded and welded on a crane

For the plate: $S_1 = \frac{20000 \times 10\ (\text{N})}{700\ (\text{mm})} = 285$ N/mm.

For the bar: $S_2 = \frac{20000 \times 10\ (\text{N})}{628\ (\text{mm})} = 318$ N/mm.

3 On the horizontal axis of Fig. 4.2, we find $S_1 = 285$ N/mm and $S_2 = 318$ N/mm, go up to cross the middle curve which is related to $\tau = 50$ MPa (the cranes are always loaded and unloaded, therefore, the force is changing between 0 and 20,000 kg. Because of that the force type is "Repeated").

4 Then, we go left to obtain the width of the welds. The minimum width for the weld of the plate and the bar is 8 and 9 mm, respectively.

A safety factor between 1.5 and 2 is usually applied to the width of welds. We apply the factor of 2:

$$z_1 = 8 \times 2 = 16 \text{ mm}$$

$$z_2 = 9 \times 2 = 18 \text{ mm}$$

Note: As shown in Table 4.2, the minimum width of the weld should correspond with the thickness of the welded plates (or diameter of the bar).

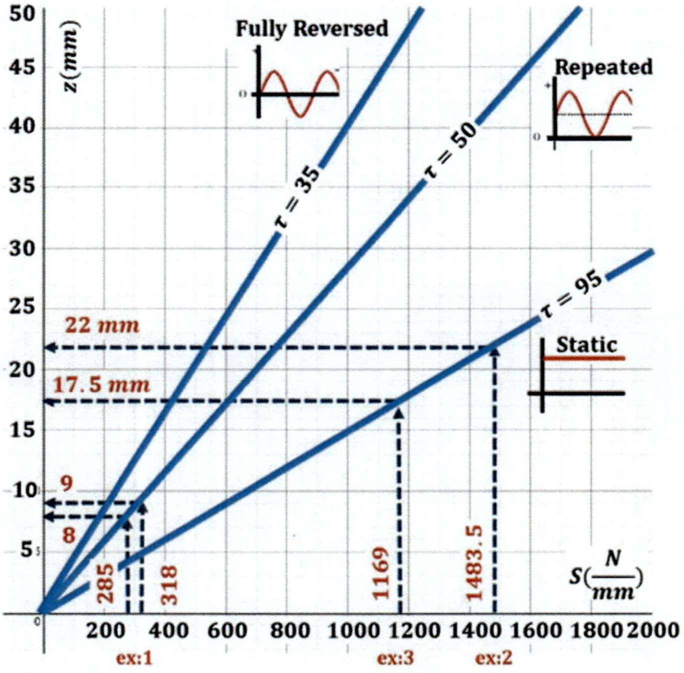

Fig. 4.2 A graph for estimation of the width of welds

Table 4.2: nMinimum width required for a weld based on the thickness of the welded elements (Hall 1961)

Thickness (t, mm)	The width of weld (z, mm)
$t \leq 10$	$z \geq 4$
$10 < t \leq 20$	$z \geq 6$
$20 < t \leq 30$	$z \geq 8$
$30 < t \leq 50$	$z \geq 10$
$50 < t \leq 150$	$z \geq 12$
$t > 150$	$z \geq 16$

In this example, the thicknesses of the plate and the bar are 30 and 100 mm, respectively. Since the minimum width of the weld for the plate and the bar should be 8 and 12 mm (Table 4.2), the obtained values for both (16 and 18 mm, respectively) are correct.

4.2.2 Bending Loading

For the cantilevered bars that are welded to a fixed support and loaded as in Fig. 4.3, first we should determine the linear stress and then estimate the width of weld using Fig. 4.2. The linear stress can be calculated using Eq. 4.3 (see Sect. 4.3).

Example 4.2 In Fig. 4.4, a plate is welded to a fixed support. A triangular stiffener is also welded under the plate. A static load of $F = 20{,}000$ kg is applied to the plate. Find the width of weld (z).

Solution

1. First, we should determine the value of Z_w. For that, we calculate n:

$$n = \frac{b}{d} = \frac{200}{150} \approx 1.3$$

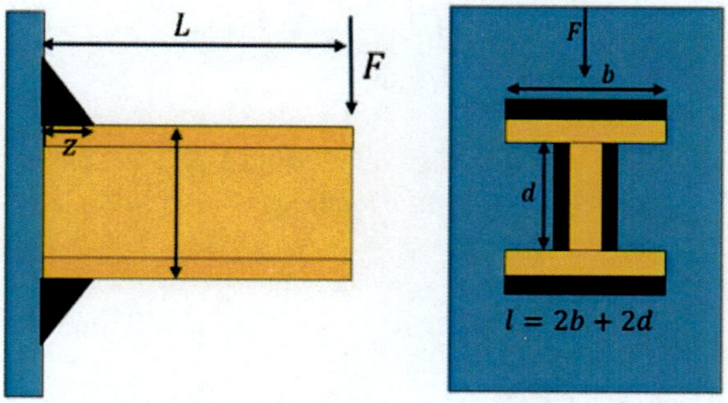

Fig. 4.3 A cantilevered bar welded fixed on a part

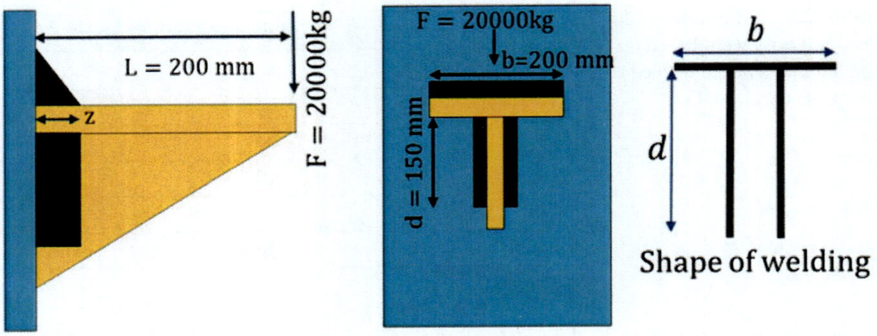

Fig. 4.4 A plate welded to a fixed support and loaded

On the left horizontal axis of Fig. 4.5, we spot $n \approx 1.3$, go up to reach the curve related to 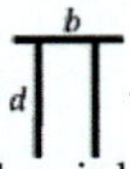, go right to cross the curve of $d = 150$ mm and then go down to reach the right horizontal axis. The estimated value of Z_w is:

$$Z_w = 28000 \text{ mm}^2$$

2. Then, we find the value of l. Again on the left horizontal axis of Fig. 4.5, we select $n \approx 1.3$, but this time we go down to cross the curve of 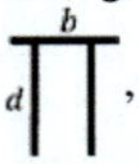, then go right to achieve the curve of $d = 150$ mm. From that point, going up to the right horizontal axis gives us the value of l:

$$l = 500 \text{ mm}$$

3. Now we can calculate S, using Eq. 4.3 (see Sect. 4.3) (here the value of force which is in kg should be multiplied by 10 to have it in N):

$$S = F \times \sqrt[2]{\left(\frac{1}{l}\right)^2 + \left(\frac{L}{Z_w}\right)^2} = (10 \times 20000) \times \sqrt[2]{\left(\frac{1}{500}\right)^2 + \left(\frac{200}{28000}\right)^2}$$
$$= 1483.5 \text{ N/mm}$$

4. To obtain z, in Fig. 4.2, we find $S = 1483.5$ N/mm on the horizontal axis, go up to the curve of *static loading* ($\tau = 95$ MPa), and go left to achieve the minimum length of weld:

$$z = 22 \text{ mm}$$

Example 4.3 Suppose that the shape of weld of Example 4.2 is as shown in Fig. 4.6 with two triangular stiffeners. Estimate the minimum width of weld.

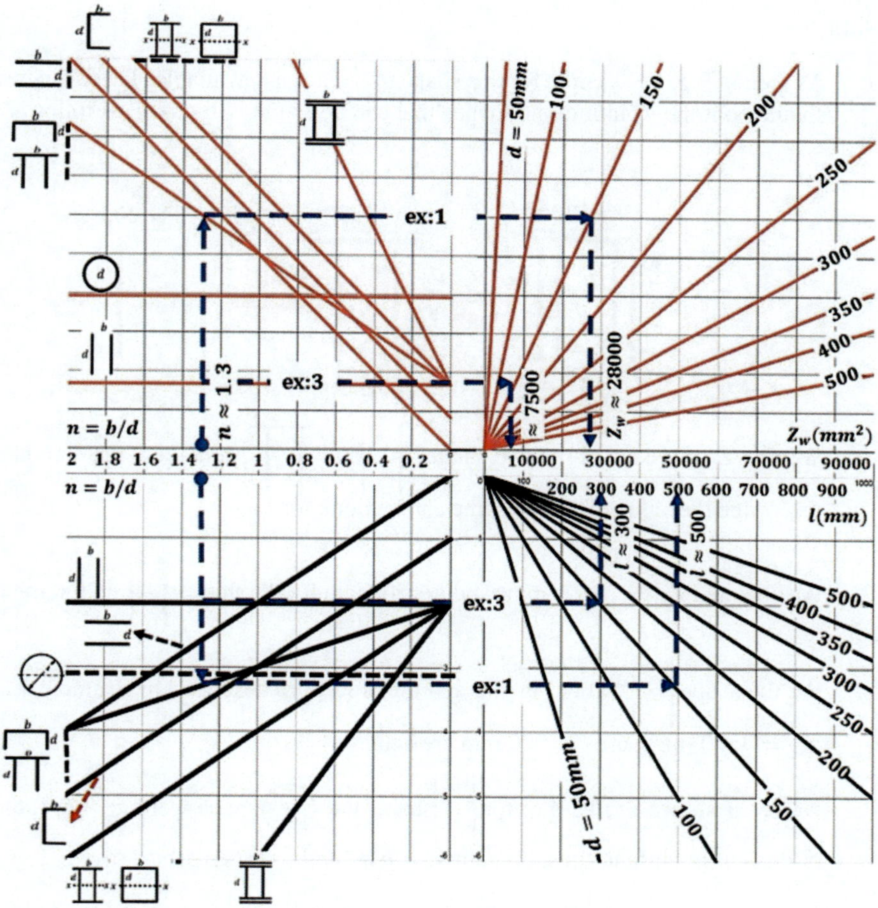

Fig. 4.5 Graphs for designing the weld of the cantilevered bars

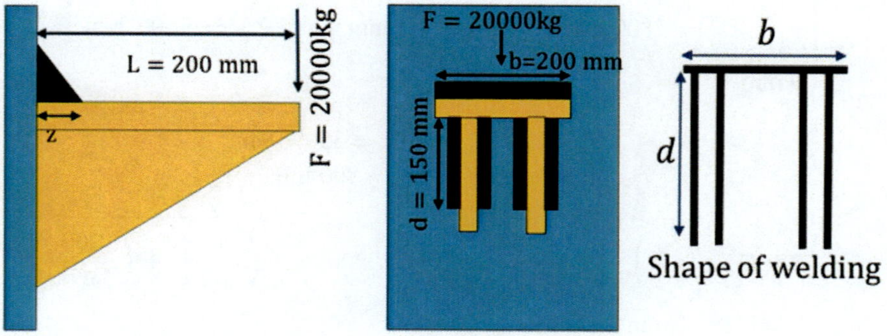

Fig. 4.6 A plate fixed to a support by two welded stiffeners (Example 4.3)

Solution

1. As seen in Fig. 4.5, a curve for this shape of weld is not available. Therefore, we should split the weld into the shapes that can be found in Fig. 4.5, as follows:

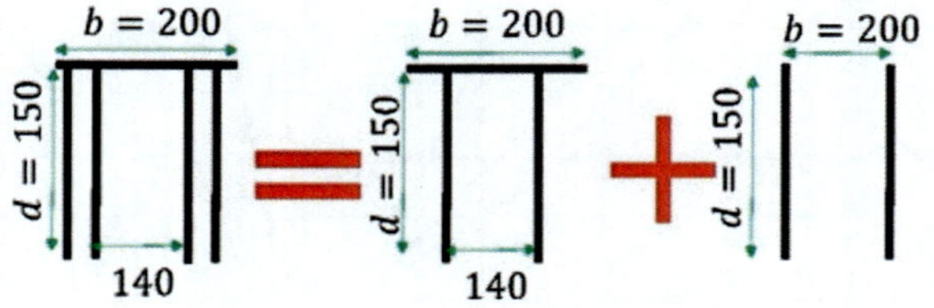

We find Z_w for each of the two shapes, 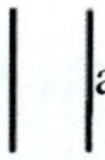 and , sum them up, and then determine the value of S. The same can be done for l.

2. We obtained $Z_w = 28000\ \text{mm}^2$ and $l = 500$ mm for the shape  in Example 4.2.

3. For the shape , using Fig. 4.5 and the procedure described in Example 4.2, $Z_w = 7500\ \text{mm}^2$ and $l = 300$ mm are achieved.

Note: As shown in Fig. 4.5, for 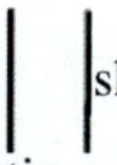 shape and  shape welding, the value of b is not important, thus, the ratio $n = b/d$ can be chosen arbitrarily.

4. Now, we sum up the obtained values of Z_w and also the values of l of the two shapes and calculate the value of S:

: $Z_w = 28000\ \text{mm}^2$, $l = 500$ mm. : $Z_w = 7500\ \text{mm}^2$, $l = 300$ mm

$$Z_w = 28000 + 7500 = 35000\ \text{mm}^2$$
$$l = 500 + 300 = 800\ \text{mm}$$

$$S = F \times \sqrt[2]{\left(\frac{1}{l}\right)^2 + \left(\frac{L}{Z_w}\right)^2} = (10 \times 20000) \times \sqrt[2]{\left(\frac{1}{800}\right)^2 + \left(\frac{200}{35000}\right)^2}$$
$$= 1169\ \text{N/mm}$$

5. Using $S = 1169$ N/mm and Fig. 4.2, $z = 17.5$ mm is obtained.

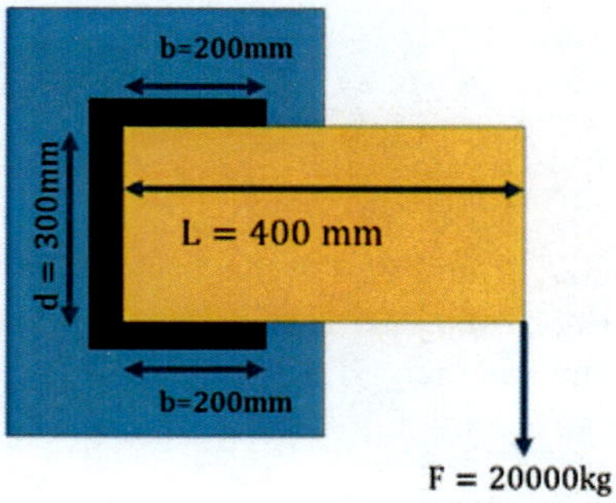

Fig. 4.7 Schematic of a plate welded to a fixed support (Example 4.4)

Example 4.4 a plate is welded to another fixed plate as shown in Fig. 4.7. Find the minimum width of weld. The applied force is static.

Solution

In the cases in which the plate is welded on another plate (like Fig. 4.7), and as a result, it is under torsional moment, the calculations are the same as the previous examples. However, to obtain Z_w, Fig. 4.8 should be used.

1. First, we calculate the length of weld, l:

$$l = 2 \times b + d = 2 \times 200 + 300 = 700 \text{ mm}$$

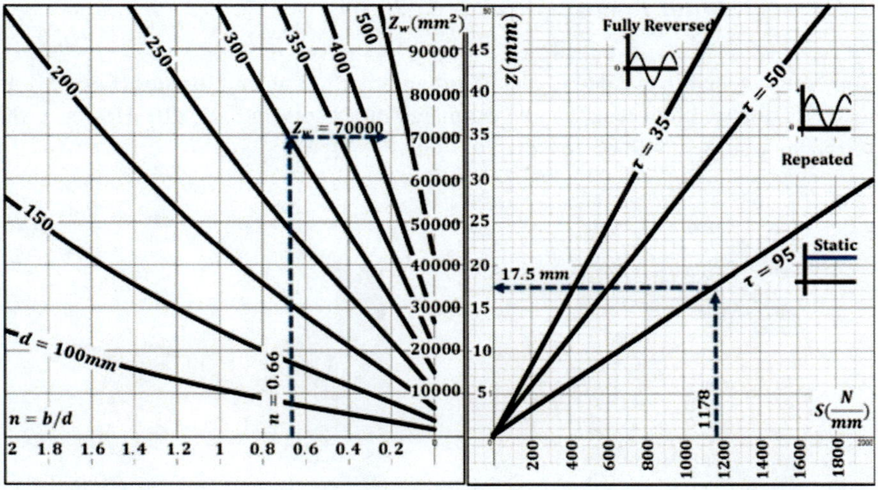

Fig. 4.8 graphs for designing the weld of the cantilevered bars under torsional loading

and n:

$$n = \frac{b}{d} = \frac{200}{300} = 0.66$$

2. Then, we determine Z_w from Fig. 4.8, that is $Z_w = 70,000\ \text{mm}^2$ (to obtain the value, from the left horizontal axis, we select $n = 0.66$, go up to intersect the curve of $d = 300$ mm, and then go right).
3. Now we can calculate S using Eq. 4.3 (see Sect. 4.3):

$$S = F \times \sqrt[2]{\left(\frac{1}{l}\right)^2 + \left(\frac{L}{Z_w}\right)^2} = (10 \times 20000) \times \sqrt[2]{\left(\frac{1}{700}\right)^2 + \left(\frac{400}{70000}\right)^2}$$
$$= 1178\ \text{N/mm}$$

4. The value of z can be obtained from Fig. 4.2 (also from Fig. 4.8) that is $z =$ 17.5 mm.

Note: If the weld width is equal to the plate thickness ($z = t$) and $b = (1 - 3)d$ (the value 3 is for high strength steels), there is no need to calculate the weld strength. Because the weld strength is higher than the plate strength. Therefore, the designer should only calculate the plate thickness. However, according to Orlov (1988), $b \geq \frac{d}{2}$ is recommended.

4.2.3 *Torsional Loading (Welded Shaft)*

Sometimes a shaft is welded to a coupling in order to transmit torques (Fig. 4.9). In this case, the width of weld can be estimated directly from Fig. 4.10. However, the following points should be considered:

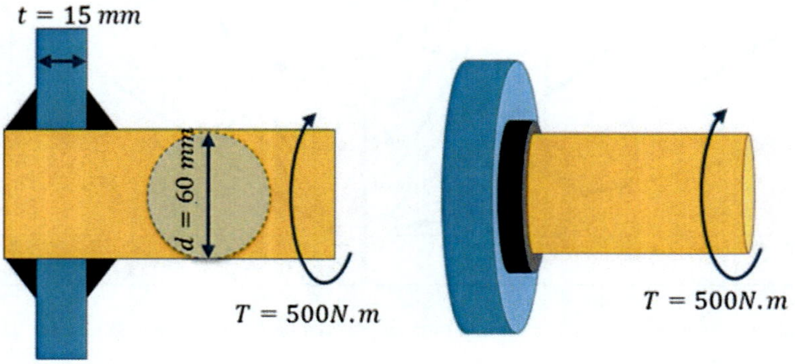

Fig. 4.9 Schematic of a shaft welded to a coupling for transmitting torque

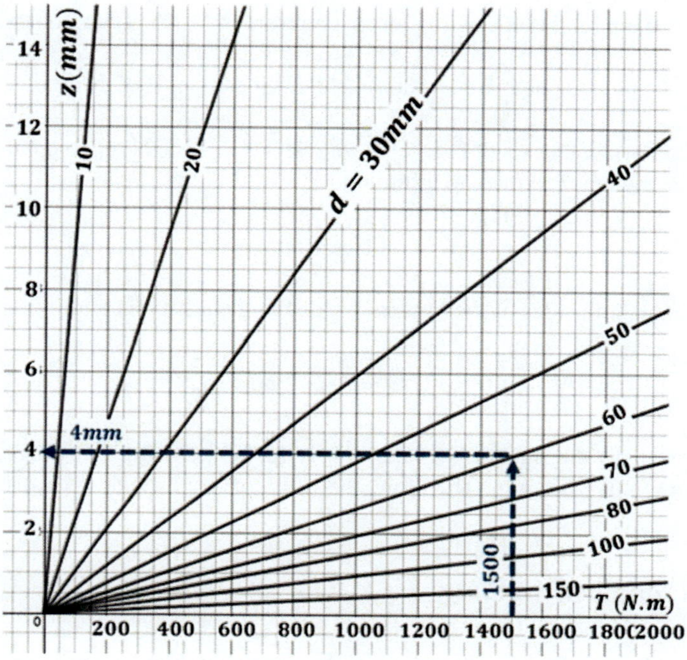

Fig. 4.10 A graph for estimating the width of weld in shafts under torsional loads

1 To obtain the principle torque (T_1), multiply the applied torque with a and b:

$$T_1 = a \times b \times T$$

Note: If the torque is *static*, , $a = 1$, if it is *repeated*, 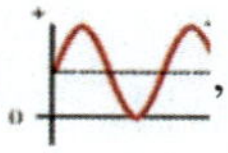, $a = 2$, and if it is *fully reversed*, 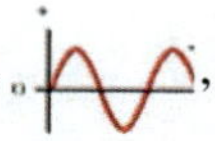, $a = 3$.

2 If bending moment is also applied to the weld (because of misalignment in shaft), multiply the torque by a factor of $b = 1.5 - 2$ (for very big bending moment, a factor of $b = 3 - 4$ can be applied).

Example 4.5 A shaft is welded to a hub (in both sides of the hub) and is supposed to transmit a torque of 500 N m. The shaft is designed to move a device back and forth. Determine the minimum width of weld.

Solution

1. First, we should determine a and b. Since the shaft moves the device back and forth, the torque is *repeated*, 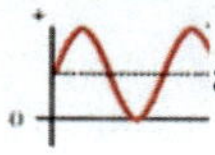 and $a = 2$. Because of manufacturing

errors, there might be bending loads, therefore, $b = 1.5$. Then, the torque is:

$$T = a \times b \times 500 = 2 \times 1.5 \times 500 = 1500 \text{ N m}$$

2. On the horizontal axis of Fig. 4.10, we select $T = 1500$ N m, go up to the curve of $d = 60$ mm, and then go left to find z:

$$z = 4 \text{ mm}$$

3. Since the both sides of the hub are welded to the shaft, the width of weld is:

$$z = \frac{4}{2} = 2 \text{ mm}$$

4. However, as the thickness of the hub is 15 mm, according to Table 4.2, the minimum of z is 6 mm. Therefore, the width of weld should be $z \geq 6$ mm.

4.2.4 Combined Loading

For the cantilevered bars (or plates) which are under combined loads (a combination of two or three of axial (tension), bending and torsional loads), as shown in Fig. 4.11, S values of each load and the total S should be determined using Eq. 4.4 (see Sect. 4.3).

Note: in narrow I-beams (bars or beams with an I-shaped cross-section) (I-Profile DIN 1025 or S-shaped) which are under bending loads (Fig. 4.12), if $z \geq 0.1\,h$, the

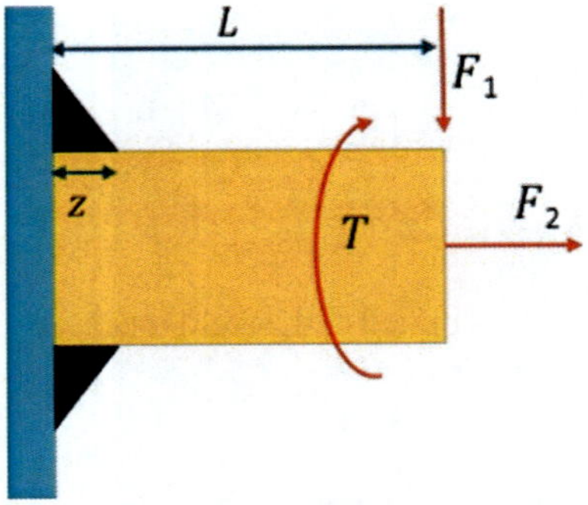

Fig. 4.11 Schematic of a cantilevered bar (or plate) under combined loads

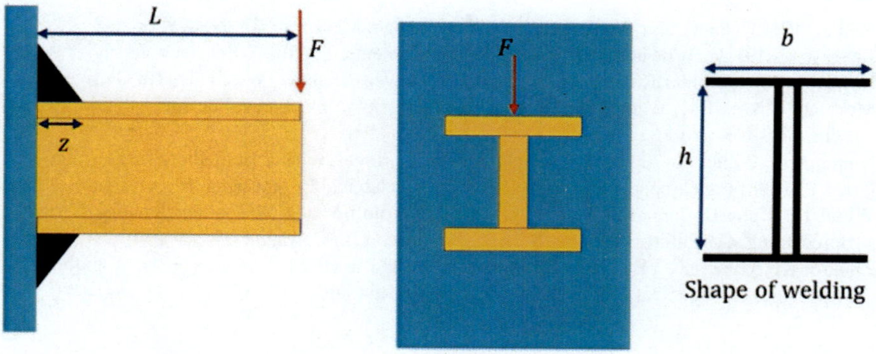

Fig. 4.12 Schematic of a cantilevered I-beam (bar with I-shaped cross-section)

weld will be stronger than the bar and there is no need for calculating z. Nevertheless, in case of other I-beams (with IPB and IPE sections), the bar is always stronger than the weld.

4.3 Equations

$$z = \frac{S}{0.707\ \tau} \tag{4.1}$$

$$S = \frac{F}{l} \tag{4.2}$$

$$\begin{cases} S = \sqrt[2]{S_1^2 + S_2^2} \\ S_1 = \frac{F}{l},\quad S_2 = \frac{FL}{Z_w} \end{cases} \rightarrow S = F \times \sqrt[2]{\left(\frac{1}{l}\right)^2 + \left(\frac{L}{Z_w}\right)^2} \tag{4.3}$$

$$S_1 = \frac{F_2}{l} + \frac{F_1 L}{Z_w},\quad S_2 = \frac{F_1}{l},\quad S_3 = \frac{Tr}{J_w}$$

$$S = \sqrt[2]{(S_1)^2 + (S_2)^2 + (S_3)^2} \tag{4.4}$$

References

Böge A (2011) Handbuch Maschinenbau: Grundlagen und Anwendungen der Maschinenbau-Technik. Vieweg+Teubner, Wiesbaden (in German)

Budynas RG, Nisbett JK (2015) Shigley's mechanical engineering design. McGraw-Hill, New York

Decker K-H, Kabus K (2014) Decker Maschinenelemente Funktion, Gestaltung und Berechnung. Carl Hanser Verlag, München (in German)

Hall A (1961) Theory and Problems of Machine Design. McGraw-Hill, New York
Jiang W (2019) Analysis and Design of Machine Elements, 1st edn. Wiley, New Jersey
Künne B (2008) Köhler/Rögnitz Maschinenteile. Vieweg+Teubner, Wiesbaden (in German)
Mott RL, Vavrek EM, Wang J (2018) Machine elements in mechanical design, 6th edn. Pearson Education, New York
Niemann G, Winter H (2003) Maschinenelemente. Springer, Berlin, Heidelberg (in German)
Orlov P (1988) Fundamentals of machine design. Mechanical Engineering, Moscow (in Russian)
Wittel H, Muhs D, Jannasch D, Voßiek J (2013) Roloff/Matek Maschinenelemente Normung, Berechnung, Gestaltung. Springer Vieweg, Wiesbaden (in German)
Zhukov KP, Gurevich YE (2014) Design of parts and assemblies of machines: a textbook for universities. Mechanical Engineering, Moscow (in Russian)

Chapter 5
Bolts

Nomenclature

A_p	Cross-sectional area of bolt head, mm^2
A_s	Effective cross-sectional area of bolt, mm^2
d	Bolt diameter, mm
F	Applied or external load, kN
F_{axial}	Axial force, kN
F_{cl}	Clamping force, kN
F_{pre}	Preload or initial tension due to tightening, kN
F_{shear}	Shear force, kN
F_{total}	Total load on bolt, kN
F_z	Preload force loss after assembly, kN
M	Tightening torque, N m
p_{per}	Allowable pressure of the connected elements, MPa
R_e	Yield strength of bolt, MPa
R_m	Ultimate tensile strength of bolt, MPa
μ	Coefficient of friction
σ_f	Fatigue strength, MPa
χ	Factor of flexibility

5.1 Introduction

A bolt is a threaded fastener that is a type of detachable joints. They are fundamental machine components as they facilitate manufacturing, assembly, disassembly, replacement and maintenance of different machine elements. Bolts are designed to

M. Yaghoubi and H. Tavakoli, *Mechanical Design of Machine Elements by Graphical Methods*, Materials Forming, Machining and Tribology,
https://doi.org/10.1007/978-3-031-04329-1_5

connect two or more components together. They are secured by tightening a nut from the opposite end of their head (Jiang 2019; Mott et al., 2018).

5.2 Design Methods

One of the most time-consuming calculations in designing machine elements is related to bolts. In this chapter, we try to simplify the calculations regarding design of bolt by applying a graphical method.

In the process of designing, first, a bolt with a specific diameter is chosen, based on the applied force. Then, it is checked to figure out whether it fulfills the required conditions for the function or not.

In this book, three methods are considered for calculations of bolt.

5.2.1 *Method I*

This method—that is based on the approaches usually used by Russian engineers (Tyunyaev et al. 2013)—is summarized as follows:

1 First step: initial estimation of the bolt diameter according to the applied force, using Eq. 5.1 (see Sect. 5.3).
2 Second step: checking the diameter based on the bearing stress of elements and the fatigue strength of bolt, using Eq. 5.2 (see Sect. 5.3).

The following examples illustrate the graphical method that we propose for performing the calculations without using the formula.

Example 5.1 A fluctuating force changing between 0 and 200 kN is applied to the cap (or flange) of a hydraulic cylinder (Fig. 5.1). Four bolts are used as fasteners. Find the diameter of the bolts. Use the bolt property class of 10.9.

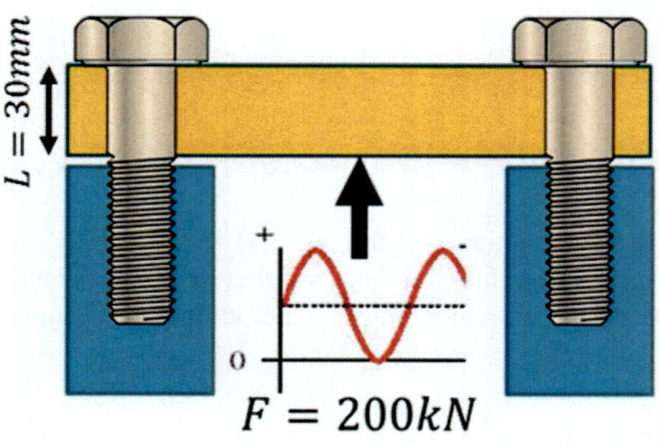

Fig. 5.1 Bolts for joining parts of a hydraulic cylinder

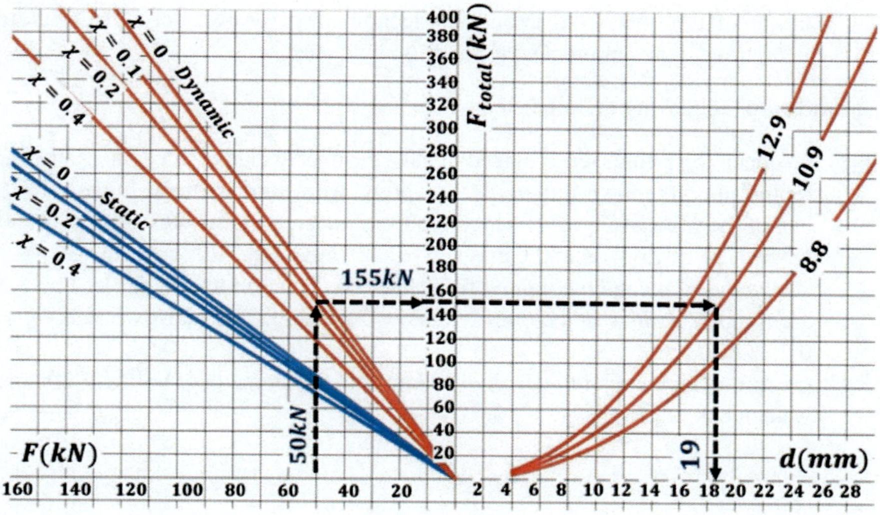

Fig. 5.2 Graphs for estimation of bolt diameter (Method I)

Solution

First step—initial estimation of the bolt diameter: Fig. 5.2 can be used for this purpose.

1. The force acting on each bolt is equal to the applied force to the flange (F) divided by number of bolts:

$$F = \frac{200}{4} = 50 \text{ kN}$$

2. Since the force is a fluctuating load, we use the curves of "*dynamic*" in Fig. 5.2. On the left horizontal axis, we select $F = 50$ kN, go up to cross the curve related to $\chi = 0.1$ (for the initial estimation, we usually select 0.1 or 0.2 for χ), and go right to obtain F_{total}:

$$F_{total} = 155 \text{ kN}$$

(We will use this force in the second step for checking the diameter.)

3. Then, we continue going right to reach the curve of the bolt property class 10.9.
4. From the point of intersection, we go down to attain the estimated diameter:

$$d = 19 \text{ mm}$$

5. We round up the estimated bolt size to the standard size:

$$d = 20 \text{ mm (M20)}$$

The calculations above are enough for technicians. However, for more accurate calculations, you can proceed to the following step.

Second step—checking the diameter

6. To check the estimation, we need to have the exact value of χ which is a factor representing the flexibility of bolt and body. We determine the value of χ from one of the plots of Fig. 5.3, which are from different references (Budynas and Nisbett 2015; Zhukov and Gurevich 2014; Autoesk Inc 2021; Orlov 1988).
7. The value of n can be obtained from Fig. 5.4. The bolt and flange of Fig. 5.1 are roughly similar to the ones with $n = 0.5$ in Fig. 5.4. Thus, we select $n = 0.5$.
8. Using the thickness of flange, $L = 30$ mm, and the diameter of bolt, $d = 20$ mm, we can calculate m:

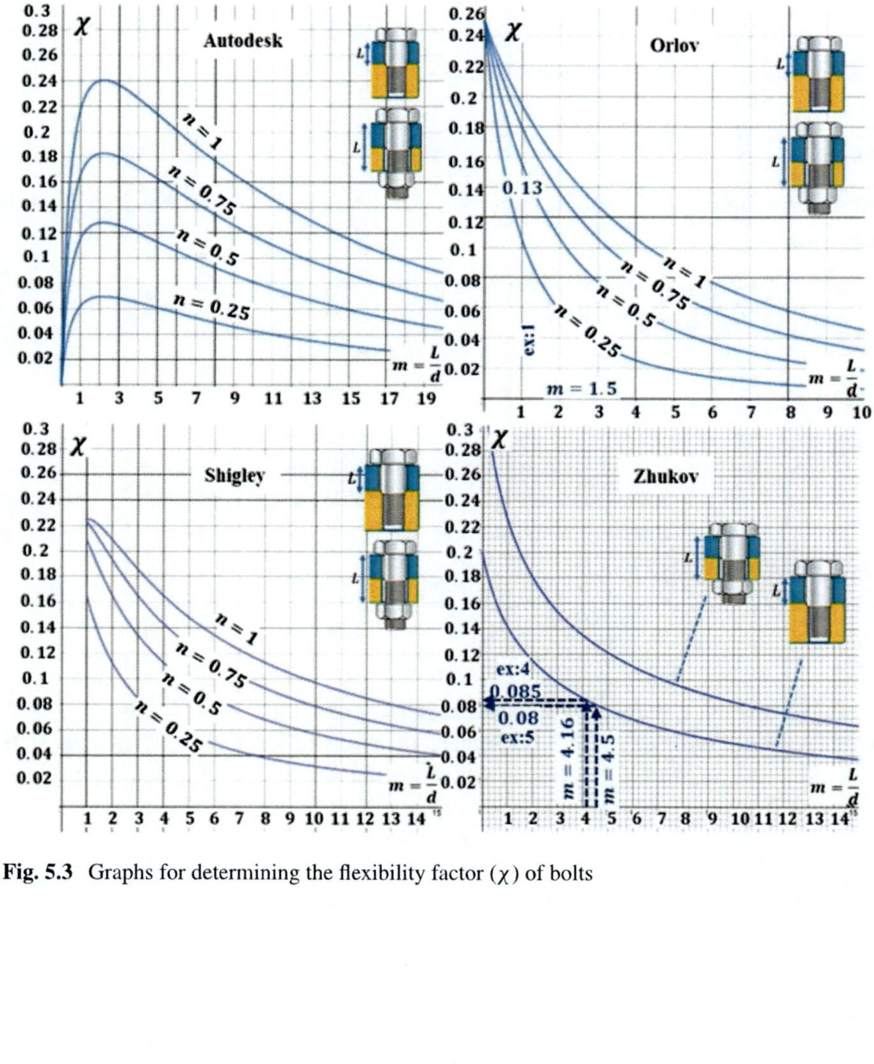

Fig. 5.3 Graphs for determining the flexibility factor (χ) of bolts

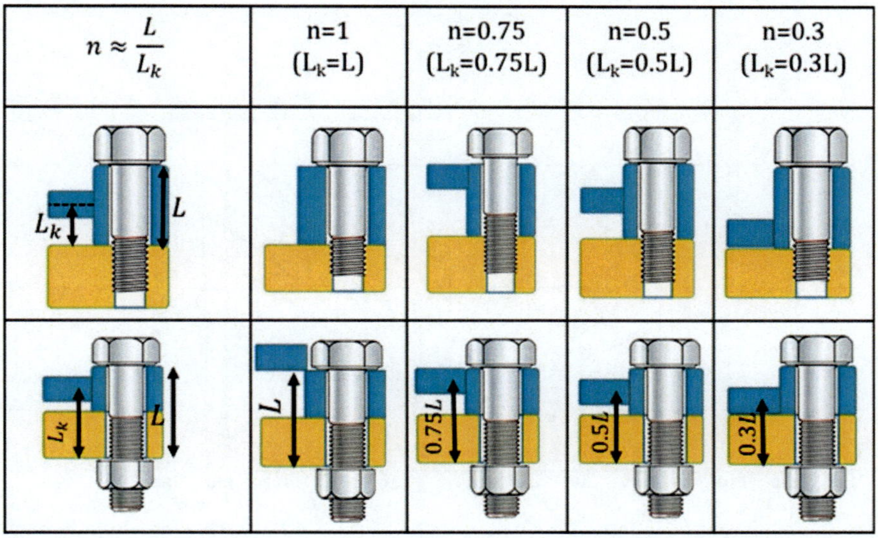

Fig. 5.4 The factor *n* for different arrangements of bolt and body

$$m = \frac{L}{d} = \frac{30}{20} = 1.5$$

9. According to $n = 0.5$ and using any of the four plots of Fig. 5.3, the value of χ can be estimated. Here, we select the plot of "*Orlov*". Thus:

$$\chi = 0.13$$

Note: if you have difficulties or doubts for finding the value of *n*, use the plot of "Zhukov" for estimating χ.

10. The bolt should be checked in terms of fatigue failure. On the left horizontal axis of Fig. 5.5, we select applied force $F = 50$ kN, go up to reach the approximate curve of $\chi = 0.13$, and then go right to reach the vertical axis and obtain *d*:

$$d = 12\,\text{mm (M12)}$$

11. Between the two values obtained for the diameter of bolt, M12 and M20, we select the bigger one (i.e. M20).

The tightening torque can be computed using Eq. 5.3 (see Sect. 5.3).

Instead of the equations above, graphs presented in Fig. 5.6 can be employed for determining the tightening torque.

Example 5.2 Estimate the tightening torque required for the bolts in Example 5.1.

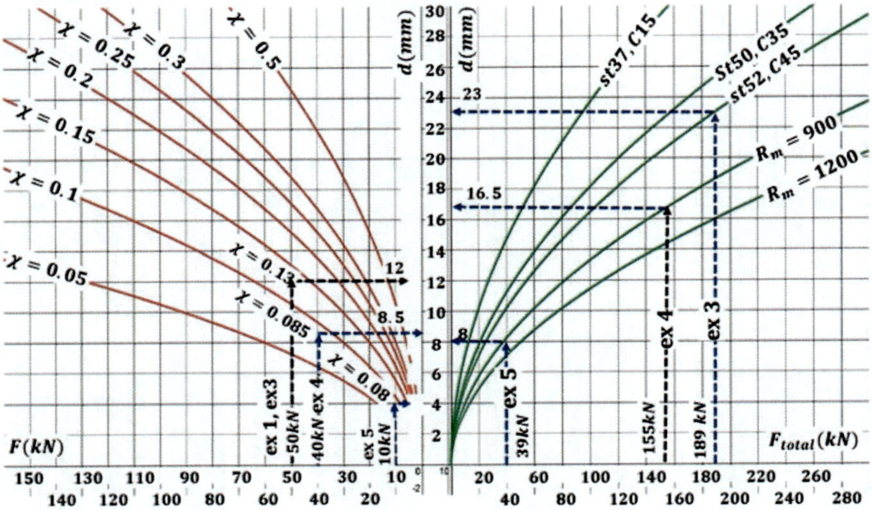

Fig. 5.5 Graphs for checking bolt diameter (in terms of fatigue failure and crushing)

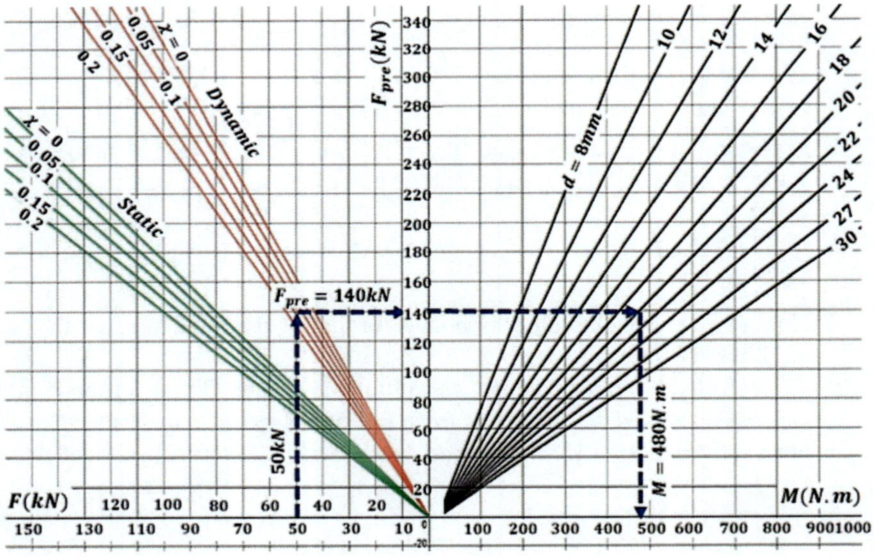

Fig. 5.6 Graphs for estimating the tightening torque of bolt (Method I)

Solution

1. On the left horizontal axis of Fig. 5.6, we select $F = 50$ kN, go up to cross the approximate curve of $\chi = 0.13$, and then go right to achieve the value of preload, F_{pre}:

$$F_{pre} = 140\,\text{kN}$$

2. We continue going right to cross the curve related to M20, and then going down to the horizontal axis gives us the tightening torque:

$$M = 480\ \text{N m}$$

3. From Fig. 5.6, a value of 140 kN obtained for the preload force; that means after tightening the bolt (with a torque equal to $M = 480$ N m), a force of 140 kN compresses the bolt and the body together.

5.2.2 *Method II*

This method is based on the DIN standards that is more applicable for technicians. According to the DIN standards, a bolt is designed so that 90% of its strength is related to tolerating the preload and the rest 10% to withstanding the applied force. While, in Method I, it is assumed that the strength of the bolt for tolerating the preload and the applied load is distributed almost equally.

The procedure of applying Method II for designing bolts is illustrated in the following examples.

Example 5.3 Solve Example 5.1 using Method II.

Solution

First step—we should find an estimated value for the diameter of bolt.

1. The force acting on each bolt is equal to the applied force to the flange (F) divided by number of bolts:

$$F = \frac{200}{4} = 50\,\text{kN}$$

2. Since the force is a fluctuating load, we use the curves of "*dynamic*" in Fig. 5.7. On the right horizontal axis, we select $F = 50$ kN, go up to cross the curve related to the bolt property class 10.9, and go left to attain the bolt diameter:

$$d = 19\,\text{mm}$$

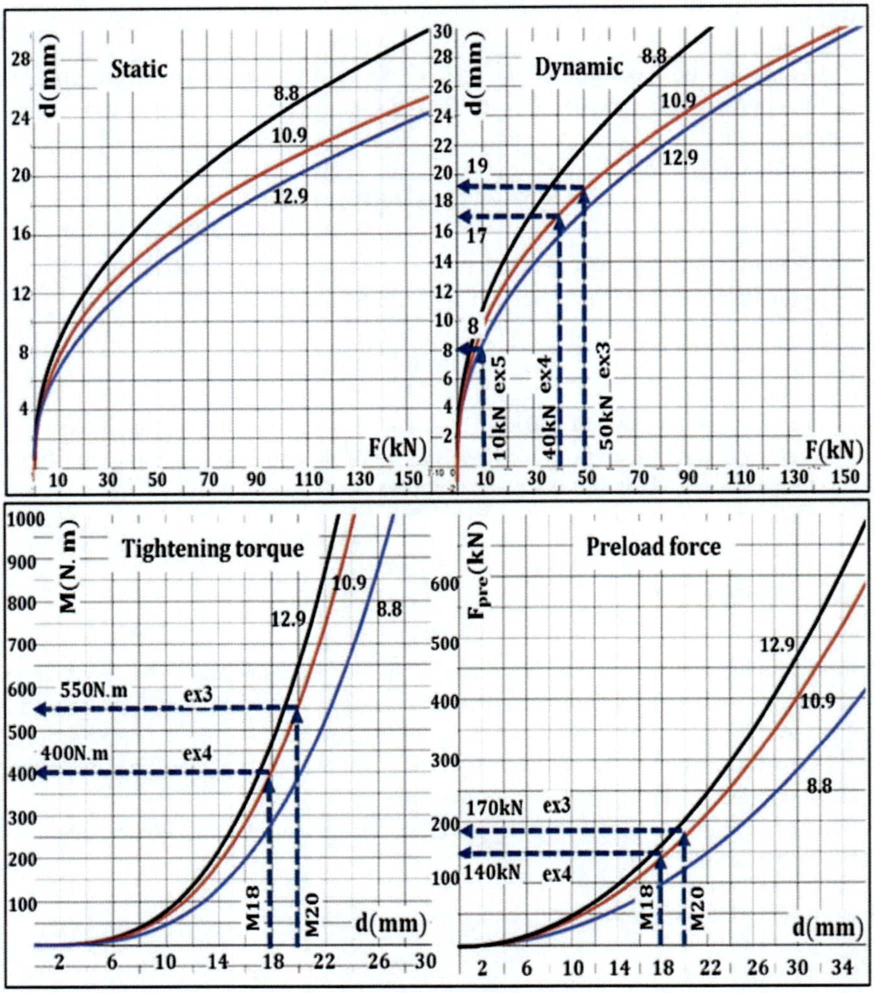

Fig. 5.7 Graphs for estimating the diameter and tightening torque of bolt (Method II)

3. We round up the estimated bolt size to the standard size:

$$d = 20\,\text{mm}\,(\text{M20})$$

4. To determine the tightening torque of bolt, we either refer to the DIN standards or use Fig. 5.7.
5. In the third quadrant (the lower left corner) of Fig. 5.7, we select $d = 20$ mm on the horizontal axis, go up to the curve of the bolt property class 10.9, and then go left to obtain M:

$$M = 550\,\text{N m}.$$

6. Also for estimating the preload force, F_{pre}, we can either use the DIN standards or employ the graphs of Fig. 5.7.
7. In the fourth quadrant (the lower right corner) of Fig. 5.7, we find $d = 20$ mm (M20), go up to cross the curve of the bolt property class 10.9, and then going left to the vertical axis results F_{pre}:

$$F_{pre} = 170\,\text{kN}$$

That means the maximum allowable force after tightening for the bolt property class of 10.9, with the size of M20 is 170 kN.

Up to this stage is enough for technicians. However, for more accurate calculations, you can proceed to the second step.

Second step—checking the diameter

The calculations above are enough for technicians and engineers. However, for checking the calculations, the following steps can be taken:

8. For checking the calculations, first we should estimate χ. In Example 5.1, the value of $\chi = 0.13$ was obtained for it.

   The bolt should be checked in terms of fatigue failure.
9. Based on Example 5.1, the minimum allowable diameter of bolt, from this point of view, is $d = 12$ mm. Therefore, M12 is a proper choice.

   In this step, we should check if the flange (or cap) of the cylinder which is made of *St52*, is crushed under the force of bolt or not (it needs washer or not). (In fact, this calculation is not necessary.)
10. We divide $F_{pre} = 170$ kN by 0.9 to obtain F_{total}:

$$F_{total} = \frac{F_{pre}}{0.9} = \frac{170}{0.9} = 189\,\text{kN}$$

11. On the right horizontal axis of Fig. 5.5, we select $F_{total} = 189$ kN, go up to cross the curve of *St52*, go left to the vertical axis and obtain $d = 23$ mm. (since this calculation was not necessary, a bolt with smaller diameter like M20 can also be selected.)

    According to the DIN standard, 10% of the strength of bolt is considered for withstanding the applied force. Thus, now we check whether the 10% of the strength of bolt can tolerate the applied force or not.
12. From Fig. 5.8, on the left horizontal axis, we select $F = 50$ kN, go up to the approximate curve related to $\chi = 0.13$, go right to reach to the curve of the property class 10.9, then going downward to the horizontal axis yields $d =$ 12 mm or M12. This means that the required diameter of bolt for tolerating $F = 50$ kN with 10% of its strength, should be greater than or equal to 12 mm.

    As a conclusion, after checking the calculations, the estimated diameter, $d = 20$ mm, is appropriate for the bolt of this example.

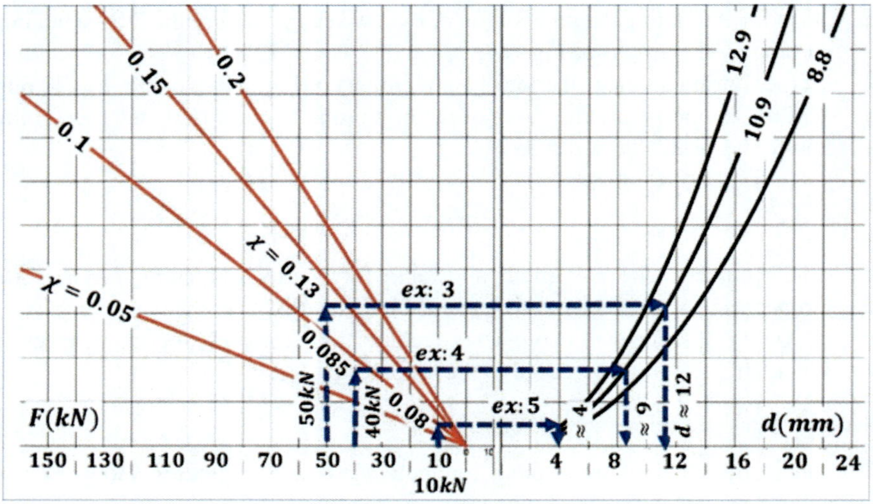

Fig. 5.8 Graphs for checking the calculations of Method II (10% of the strength of bolt)

Example 5.4 Figure 5.9 shows a piston and cylinder system. The piston is under a force of $F = 40$ kN. It is connected to the connecting rod using a bolt, as shown. To keep the connection of piston and rod sealed, they should be pressed together with a force of $F_{cl} = 15$ kN (the clamping force). Find the size (diameter) of bolt. (Use the bolt property class of 10.9).

Solution

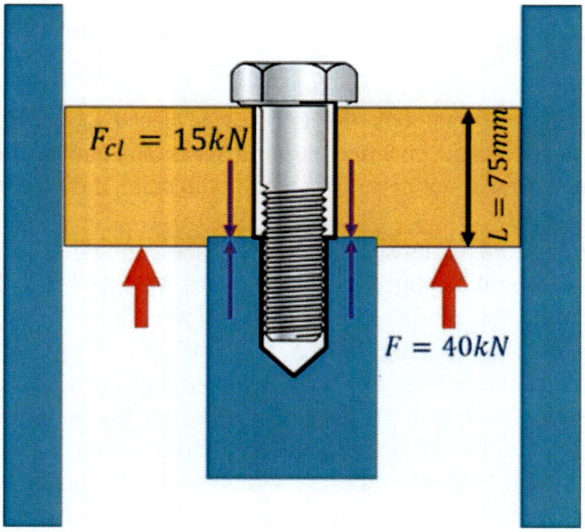

Fig. 5.9 Schematic of a piston and cylinder system

We solve this example using Method II.

1. In the first quadrant (the upper right corner) of Fig. 5.7, we find the estimated diameter of bolt based on $F = 40$ kN and the bolt property class of 10.9, that is $d = 17$ mm. Thus, M18 is selected. For this example, the curve related to *dynamic* forces is used, because the cylinder is moving and therefore the applied force is varying.
2. From the third and fourth quadrants of Fig. 5.7, using $d = 18$ mm and the bolt class 10.9, we obtain the tightening torque, $M = 400$ N m, and the preload force, $F_{pre} = 140$ kN.

   Now, we check the calculations.
3. Since for this example it is not easy to find the value of n from Fig. 5.4, we use the Zhukov's method (the fourth quadrant of Fig. 5.3) to estimate the value of χ. For that, we need the value of m:

$$m = \frac{L}{d} = \frac{75}{18} = 4.16$$

   Considering $m = 4.16$, $\chi = 0.085$ is obtained.

   Now we check the bolt design in terms of fatigue failure and crushing, as well as if it can tolerate the applied force by 10% of its strength. (Suppose that the piston is made of quenched and tempered ck45 with an ultimate tensile strength of $R_m = 900$ MPa.)
4. The value of F_{total} is:

$$F_{total} = \frac{F_{pre}}{0.9} = \frac{140}{0.9} = 155 \text{ kN}$$

5. Using Fig. 5.5, in terms of crushing, the minimum diameter should be 16.5 mm, and in the sense of fatigue failure, it should be minimum 8.5 mm. In addition, based on Fig. 5.8, for tolerating the applied force by 10% of strength of bolt, the diameter should be at least 9 mm. Therefore, M18 is a proper choice for this example.

   Note: since the difference between estimated diameter, M18, and the two checked diameters, 8.5 and 9 mm, is too big, we can select M16 or M14 too (with ignoring crushing).
6. The clamping force required for sealing the connection is $F_{cl} = 15$ kN. Here, we examine that with choosing M18, the connection remains sealed or not. For this purpose, we use Eq. 5.4 (see Sect. 5.3) to calculate the clamping force.

   Thus, F_{cl} is:

$$F_{cl} = F_{pre} \times (1 - \chi) - F_z = 140 \times (1 - 0.085) - 0.15 \times 40 = 122.1 \text{ kN}$$

   The obtained value is much greater than the required clamping force of 15 kN. Therefore, the sealing is surely provided.

Note: to reduce the clamping force, the preload force should be decreased. This can be done by loosening the bolt. By placing $F_{cl} = 15$ kN in Eq. 5.4 (see Sect. 5.3), we have:

$$F_{cl} = F_{pre} \times (1 - \chi) - F_z \rightarrow$$

$$15 = F_{pre} \times (1 - 0.085) - 0.15 \times 40 \rightarrow F_{pre} = 23\,\text{kN}$$

7. Now we can estimate the tightening torque from Fig. 5.6 using $F_{pre} = 23$ kN and $d = 18$ mm:

$$M \approx 70\,\text{N m}$$

Example 5.5 The connecting rod of a piston system is joined to its cap using two waisted bolts as shown in Fig. 5.10. A force of 10 kN is applied to each bolt. Consider the bolt property class of 12.9 and $L = 45$ mm. Find the size (diameter) of bolt.

Solution

1. First, from Fig. 5.7, using $F = 10$ kN (*dynamic*) and the bolt class of 12.9, $d =$ 8 mm (M8) is obtained.

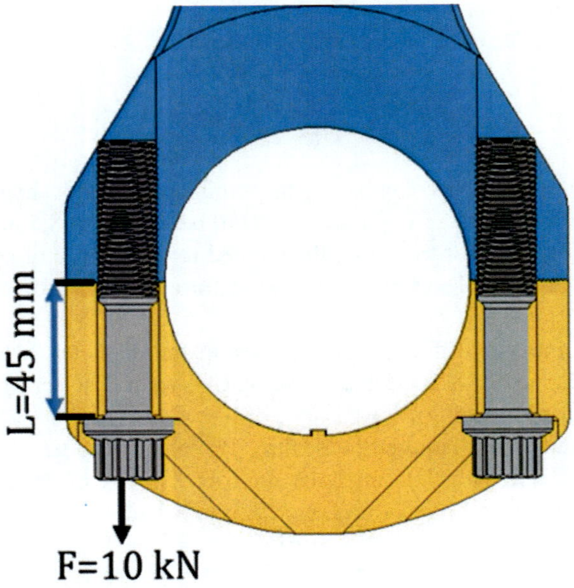

Fig. 5.10 Joining a connecting rod of a piston to its cap using two waisted bolts

Since the bolts are waisted, we can use bolts with one or two sizes bigger or use the graph of Fig. 5.11. Thus, the estimated diameter is 10 mm (M10).

2. From Fig. 5.12, we can find the preload and the tightening torque of the bolts:

$$F_{pre} = 35\,\text{kN}$$

$$M = 50\,\text{N m}$$

3. We estimate the value of χ using Fig. 5.3 and $\frac{L}{d} = \frac{45}{10} = 4.5$.
4. Since for this example it is not easy to find n from Fig. 5.4, we use the Zhukov's method (the fourth quadrant of Fig. 5.3). Thus, $\chi = 0.08$ is achieved.
5. Now we check the calculation is terms of the crushing of piston (which is made of quenched and tempered ck45 with an ultimate tensile strength of $R_m = 900$ MPa) of bolt and the fatigue failure from Fig. 5.5 and in the sense of

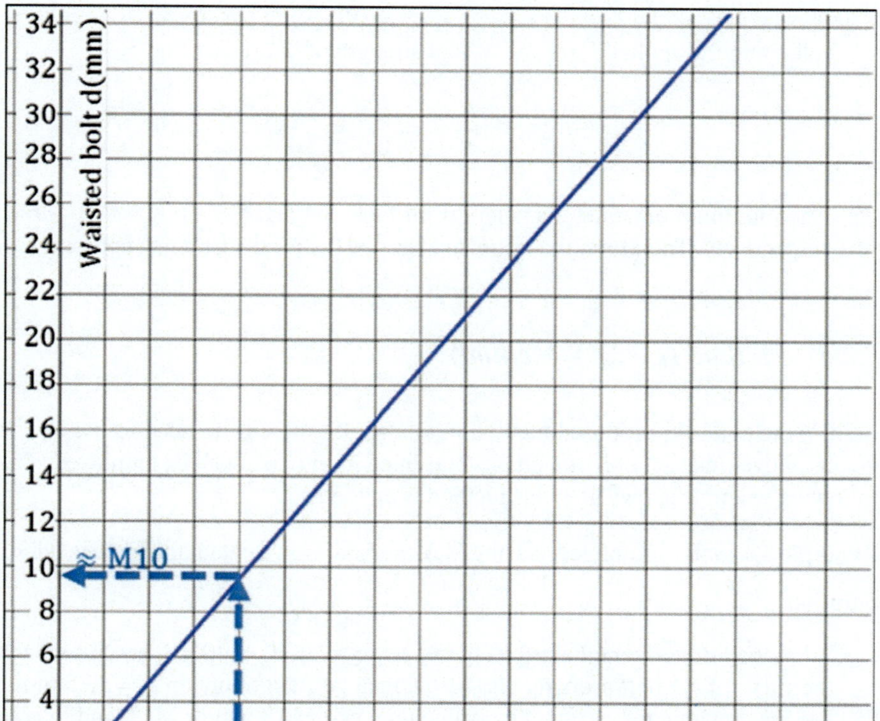

Fig. 5.11 Relationship between diameter of waisted bolts and that of shank bolts

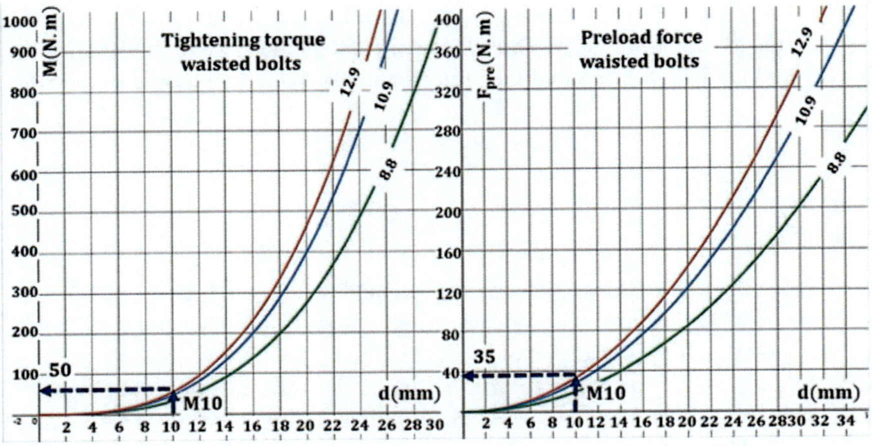

Fig. 5.12 Graphs for estimating the preload ant tightening torque of waisted bolts

tolerating the applied force by 10% of strength of bolt using Fig. 5.8.

$$F_{total} = \frac{F_{pre}}{0.9} = \frac{35}{0.9} = 39\,\text{kN}$$

The minimum obtained diameters from Figs. 5.5 and 5.8 are 8, 4 and 4 mm, respectively. Therefore, the estimated size (M10) for the bolts is appropriate.

5.2.3 *Method III (M.Y Method)*

In M.Y method (our proposed method in this book), there is no need to use all the graphs above. Instead, only the graphs presented in Fig. 5.13 can be employed. The following example shows how this method is applied.

Example 5.6 Solve Example 5.4 using M.Y method. How about if the force is static?

Solution

1 From the left horizontal axis of Fig. 5.13, we select $F = 40$ kN, go up to reach the curve of the bolt property class 10.9, then go right to obtain $d = 16.5$ mm.
2 For the dynamic force, round the value up, i.e. $d = 18$ mm or M18, and in case of the static force, round it down, i.e. $d = 16$ mm or M16.
3 On the vertical axis of Fig. 5.13, we find $d = 16$ and 18 mm, go right to reach the curve related to the bolt class 10.9, and then going down to the horizontal axis gives us the tightening torque of the dynamic and static forces, 400 and 280 N m, respectively.

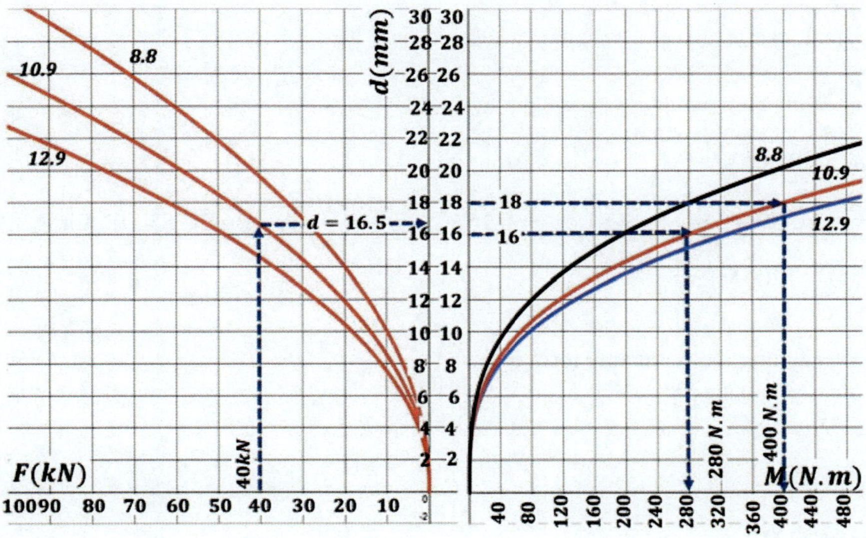

Fig. 5.13 graphs of M.Y method for designing bolt

4 To be on the safe side, reduce the tightening torques by 5–10%. Thus, the final tightening torque may be 360 and 250 N m for dynamic and static forces, respectively.

Note: If shear (F_{shear}) and axial forces (F_{axial}) are applied to the bolt at same time, for design, consider the equivalent load as external force which is: $F \approx \sqrt{(F_{axial})^2 + 4(F_{shear})^2}$.

5.3 Equations

Method I

First step: initial estimation of the bolt diameter

$$\begin{aligned} F_{total} &= v(1-\chi)F + \chi F \\ \frac{F_{total}}{R_e/1.2} &= A_s \\ A_s &= \frac{\pi}{4}(0.87d - 0.4288)^2 \\ \mathrm{v} &= \begin{cases} 1.75 & \text{(for static loads)} \\ 3.25 & \text{(for dynamic loads)} \end{cases} \end{aligned} \quad (5.1)$$

Second step: checking the estimated diameter

$$\frac{\chi F}{2A_s} = \frac{\sigma_f}{1.2}$$
$$\frac{F_{total}}{A_p} = p_{per}$$
$$A_p \approx 0.8063d^{1.9033}$$
$$\sigma_f = 0.85 \times \left(\frac{150}{d} + 45\right) \text{ or } = 0.15R_m \times \frac{25+d}{25+3d} \text{ or } = 65 - 85\,\text{MP}_\text{a} \text{ for } R_m = 600 - 1200\,\text{MPa} \tag{5.2}$$

Determining the tightening torque

$$M = 0.17F_{pre}\,d$$
$$F_{pre} = v(1-\chi)F \tag{5.3}$$

Method II

First step: initial estimation of the bolt diameter.

Bolt property class	Static load	Dynamic load
8.8	$F = 0.0665d^{2.2962}$	$F = 0.042d^{2.2954}$
10.9	$F = 0.0992d^{2.2753}$	$F = 0.0628d^{2.2745}$
12.9	$F = 0.1277d^{2.2435}$	$F = 0.0808d^{2.2432}$

Second step: checking the estimated diameter

$$F_{cl} = F_{pre}(1-\chi) - F_z$$
$$F_z \approx 0.15F \tag{5.4}$$

$$\frac{\chi F}{2A_s} = \frac{\sigma_f}{1.2}$$
$$\frac{F_{total}}{A_p} = p_{per}$$
$$\chi F \le 0.1R_eA_s \tag{5.5}$$

Preload force and tightening torque according to DIN

	Shank bolts		*Waisted bolts*	
Bolt property class	*Preload*	*Tightening torque in $\mu = 0.12$*	*Preload*	*Tightening torque in $\mu = 0.12$*
8.8	$F = 212.27d^{2.1149}$	$M = 0.0383d3.0791$	$F = 137.9d^{2.1436}$	$M = 0.0255d^{3.0996}$
10.9	$F = 323.58d^{2.0953}$	$M = 0.0596d3.0505$	$F = 209.7d^{2.1258}$	$M = 0.0383d^{3.0866}$
12.9	$F = 381.58d^{2.0922}$	$M = 0.0696d3.0513$	$F = 247.1d^{2.1232}$	$M = 0.0448d^{3.0844}$

If shear (F_{shear}) and axial forces (F_{axial}) are applied to the bolt at the same time

$$\mathrm{F} \approx \sqrt{(F_{axial})^2 + 4(F_{shear})^2} \tag{5.6}$$

References

Autodesk Inc. (2021) Engineer's Handbook. Retrieved from Autodesk Inventor support and learning center: https://knowledge.autodesk.com/

Budynas RG, Nisbett JK (2015) Shigley's Mechanical Engineering Design. McGraw-Hill, New York

Jiang W (2019) Analysis and design of machine elements, 1st edn. Wiley, New Jersey

Mott RL, Vavrek EM, Wang J (2018) Machine elements in mechanical design, 6th edn. Pearson Education, New York

Orlov P (1988) Fundamentals of Machine Design. Mechanical engineering, Moscow (in Russian)

Tyunyaev AV, Zvezdakov VP, Wagner VA (2013) Machine parts: textbook. Lan Electronic Library, Moscow (in Russian)

Zhukov KP, Gurevich YE (2014) Design of parts and assemblies of machines: a textbook for universities. Mechanical Engineering, Moscow (in Russian)

Chapter 6
Power Screws

Nomenclature

d	Screw diameter, mm
d_2	Mean diameter of screw, mm
d_3	Minimum diameter of screw, mm
F	Applied tension/compression force, kN
F_1	Allowable tension and compression force of screw, kN
F_2	Allowable force on nut, kN
F_3	Allowable buckling force, kN
L	Screw length, mm
l_n	Nut length, mm
p_p	Allowable thread pressure of nut, MPa
R_e	Yield strength of screw, MPa
R_m	Ultimate tensile strength of screw, MPa
S	Buckling safety factor
T	Torque required for turning screw, MPa
μ	Coefficient of friction
σ_p	Allowable stress of screw, MPa
σ_t	Equivalent stress, MPa

6.1 Introduction

Power screws are elements used for converting rotary motion to linear motion in machines. Applications include screw-type jacks, tensile and compression testing machines, home garbage compactors, C-clamps, vises, the lead screws of lathes, etc. There are three forms of power screw threads: the square thread, the Acme thread,

M. Yaghoubi and H. Tavakoli, *Mechanical Design of Machine Elements by Graphical Methods*, Materials Forming, Machining and Tribology,
https://doi.org/10.1007/978-3-031-04329-1_6

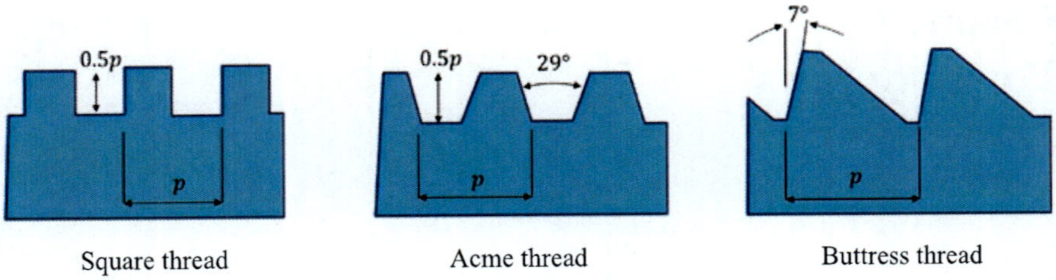

Fig. 6.1 Different forms of power screw threads

and the buttress thread (Fig. 6.1). Power screws may also be either single-thread or double-thread or triple-thread. A strong disadvantage of power screws can be their relatively low efficiency. However, using higher pitch designs (multiple-thread) can lead to higher efficiencies (Budynas and Nisbett 2015; Collins et al. 2009; Juvinall and Marshek 2011; Mott et al. 2018).

In this book, we only consider power screws with the Acme threads and provide unique graphical methods for designing them.

6.2 Design Methods

Design of power screws includes determining the diameter and pitch based on the applied force. They are two important characteristics of power screws. Table 6.1 presents the standard diameters of power screws together with their corresponding pitches (Decker and Kabus 2014).

Here, we introduce two methods for the design.

6.2.1 Method I

In this method, Eqs. 6.1 and 6.2 (see Sect. 6.3) are used (the coefficient of friction (μ) between the thread and nut is assumed to be 0.15).

If the screw is under compression loading, buckling can occur. In this case, Eqs. 6.3 and 6.4 (see Sect. 6.3) can be applied for the design, which consider this matter.

In addition, the torque required for turning the screw can be calculated from Eq. 6.5 (see Sect. 6.3).

However, instead of using the equations, the graphical method proposed in this book can be employed. We describe the use of the method in the following examples.

Example 6.1 A burden of 10 tons is supposed to be lifted by a screw-type jack, up to $L = 1000$ mm (Fig. 6.2). The power screw and nut are made of ck45 and steel, respectively. The load is applied just in one-way (compression) to the screw. The screw is a double-thread type. Find the diameter and pitch of the power screw.

Table 6.1 The standard diameter (d) and pitch (p) of power screws (Decker and Kabus 2014)

d (mm)	8	10	12	16	20	24	28	32	36	40	44	48	52	60	65	70	75	80	85	90	95	100	110	120
p (mm)	1.5	2	3	4	4	5	5	6	6	7	7	8	8	9	10	10	10	10	12	12	12	12	12	14

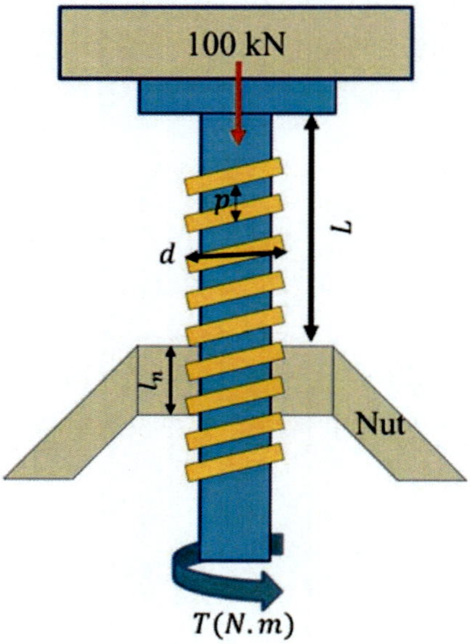

Fig. 6.2 Schematic of a power screw used for lifting a burden (by pushing it upwards)

Solution

1. First, we estimate the diameter based on the strength of screw, using Fig. 6.3. To this aim, on the left horizontal axis, we select $F = 100$ kN(10 tons $\approx$ 100 kN), and go up to the curve of ck45.

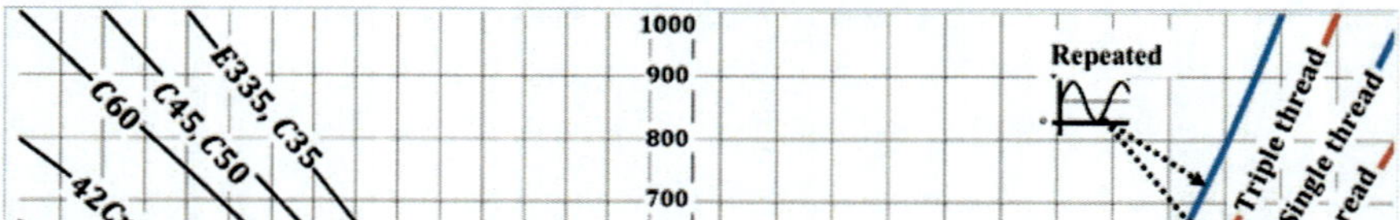

d	8	10	12	16	20	24	28	32	36	40	44	48	52	60	65	70	75	80	85	90	95	100	110	120
P	1.5	2	3	4	4	5	5	6	6	7	7	8	8	9	10	10	10	10	12	12	12	12	12	14

Fig. 6.3 Graphs for estimating the diameter of power screws according to the strength of screw

2. As the load is *one-way*, we should choose the curve related to *repeated* load. Thus, we go left to reach the approximate curve related to both *repeated* 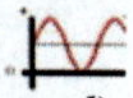 and *double-thread* (between the curves of *single-thread* and *triple-thread*). The corresponding value on the horizontal axis would be the diameter:

$$d = 45 \text{ mm}$$

3. Since $d = 45$ mm is not a standard diameter for power screws (Table 6.1), we round it up and select $d = 48$ mm from Table 6.1 or blow the horizontal axis of Fig. 6.3. Therefore, the screw size is:

$$d \times P = 48 \times 8 \text{ mm}$$

4. Here, we estimate the diameter according to the strength of nut, using Fig. 6.4. We find $F = 100$ kN on the horizontal axis, go up to the curve of *steel* (the material of nut), and then go left to obtain the diameter on the vertical axis:

$$d = 59 \text{ mm}$$

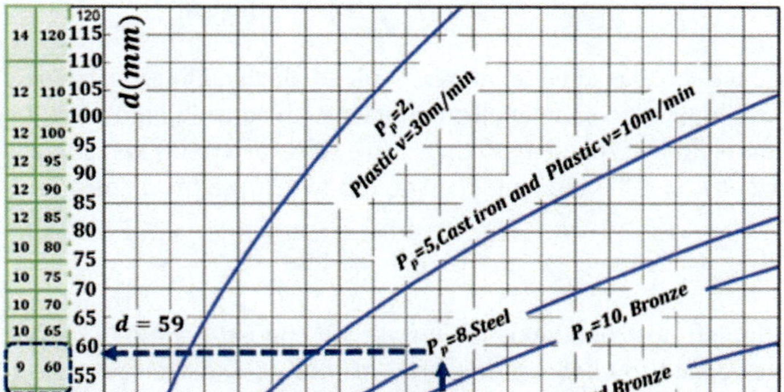

Fig. 6.4 Graphs for estimating the diameter of power screws according to the strength of nut

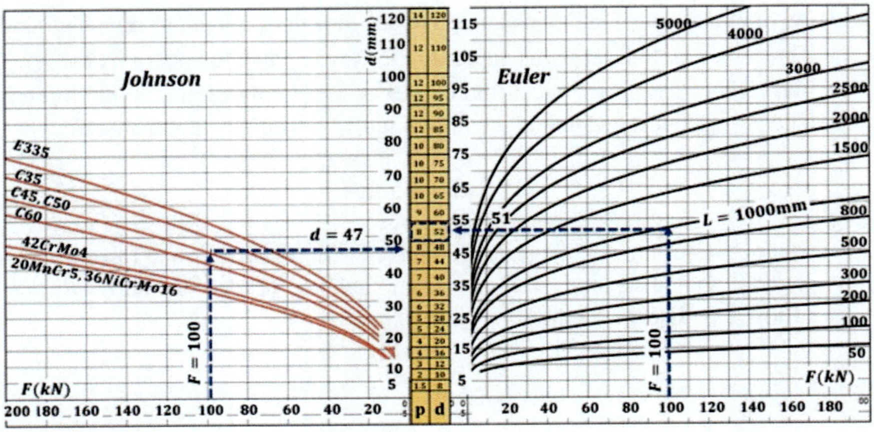

Fig. 6.5 Graphs for estimating the diameter of power screws with considering buckling

We round it up to the standard diameter of 60 mm from the table next to the vertical axis or Table 6.1. Thus, the size of the screw is:

$$d \times P = 60 \times 9 \text{ mm}$$

5. Since the screw is under compression load, the buckling can occur. Therefore, we use Fig. 6.5 to estimate the diameter with considering this matter.
6. On the right horizontal axis (*Euler*), we select $F = 100$ kN, go up to cross the curve of $L = 1000$ mm, and then go left to achieve the diameter:

$$d = 51 \text{ mm}$$

7. On the left horizontal axis (*Johnson*), we choose $F = 100$ kN, go up to the curve related to ck45 (the material of screw), and then go right to find the diameter on the vertical axis:

$$d = 47 \text{ mm}.$$

8. Between the two obtained values, we select the bigger one, $d = 51$ mm. If the screw was single-thread, we round it up to the standard diameter of 52 mm as shown in Fig. 6.5. However, to obtain the corresponding diameter of double-thread and triple-thread screws, Fig. 6.6 should be used.
9. As shown, the corresponding diameter for the double-thread screw is 55 mm. We round it up to the standard value, $d = 60$ mm. Therefore, the size of screw would be:

$$d \times P = 60 \times 9 \text{ mm}$$

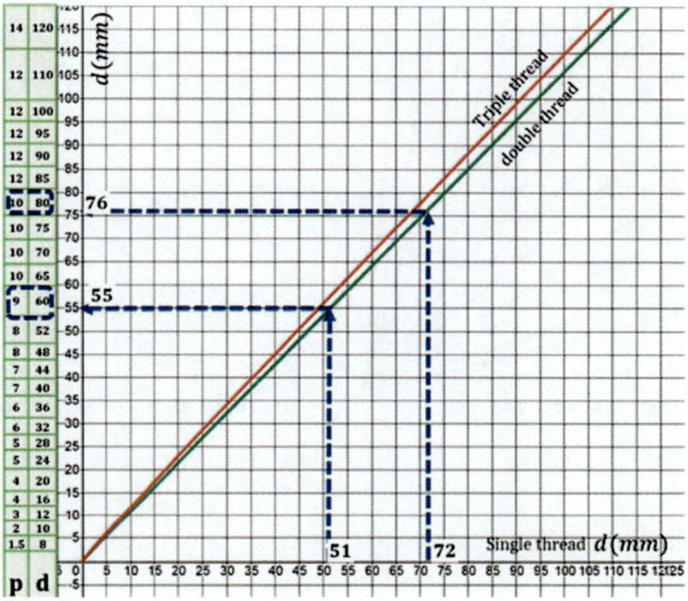

Fig. 6.6 Relationships between the diameters of single-thread and both double-thread and triple-thread screws

10. In conclusion, considering the three conditions above, we select the bigger one, i.e. $d \times P = 60 \times 9$ mm.

    Thus, the length of nut is:

$$l_n = 2.5d = 2.5 \times 60 = 150 \text{ mm}$$

11. We can show the characteristics of the screw with this format: $d \times (n \times \text{pitch})P(\text{pitch})$ in which n is the number of threads. For this example, as it is a double-thread screw, we can show the size like $60 \times 18P9$.

Example 6.2 Find the required torque for turning the screw of Example 6.1.

Solution

To determine the torque required for turning power screws, Fig. 6.7 can be employed. For the purpose of this example: we select $d = 60$ mm on the right horizontal axis, go up to the curve of *double-thread*, then go right to cross the curve related to $F = 100$ kN, and find the corresponding point on the left horizontal axis that is $T = 710 \text{ N} \cdot \text{m}$.

Therefore, a torque of 710 N·m is required to turn the power screw.

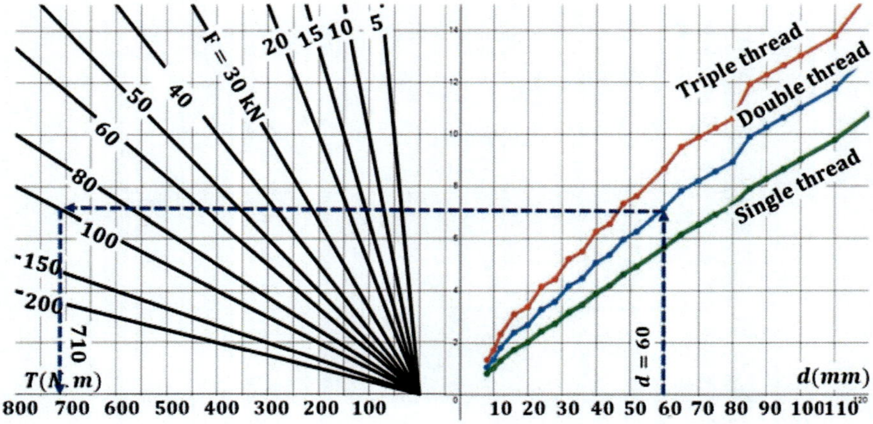

Fig. 6.7 Graphs for estimating the required torque for turning power screws

6.2.2 Method II (M.Y Method)

We propose a simple method, named as M.Y method, for designing power screws. In this method, for determining the diameter of screw, we just need to use Figs. 6.6 and 6.8. It is assumed that the material of screw is ck45 and nut is made of steel $\left(p_p = 8 \text{ MPa}\right)$ or stronger than them. In addition, the nut length is $l_n = 2.5d$. The following examples illustrate the use of the method.

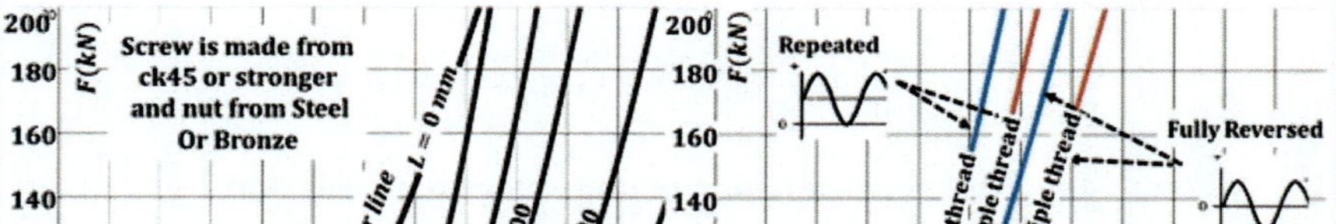

d	8	10	12	16	20	24	28	32	36	40	44	48	52	60	65	70	75	80	85	90	95	100	110	120
P	1.5	2	3	4	4	5	5	6	6	7	7	8	8	9	10	10	10	10	12	12	12	12	12	14

Fig. 6.8 Graphs for estimating the diameter of power screws (M.Y method)

Example 6.3 Solve Example 6.1 using M.Y method. Assume that the screw length is $L = 2000$.

Solution

1. In Fig. 6.8, we select $F = 100$ kN on the vertial axis of the left side and go right to cross the curves of "*Border line*" and $L = 2000$ mm; then going down to the horizontal axis gives $d = 59$ mm and $d = 72$ mm.
2. Between the two diameters, we select the bigger one, i.e. $d = 72$ mm. As the screw is double-thread, using Fig. 6.6, the corresponding diameter of 76 mm is obtained (if you do not want to use Fig. 6.6, simply select a screw with standard diameter of one or two size bigger).
3. Again in Fig. 6.8, we find $F = 100$ kN on the vertial axis of the right side, go right to reach the approximate curve of *double-thread* (between *single-thread* and *triple-thread*), and go down to achieve $d = 45$ mm.
4. Now between $d = 76$ mm and $d = 45$ mm, we choose the bigger one. According to Table 6.1 or the table in Fig. 6.8 and with rounding up the diameter of 76 mm, $d = 80$ mm is obtained.
5. Therefore, the characteristics of the power screw should be $80 \times 20P10$. In addition, the length of nut is $l_n = 2.5 \times 80 = 200$ mm.

Example 6.4 Suppose that the power screw in Example 6.1 is under tension loading (Fig. 6.9). Solve it using M.Y method.

Solution

1. Since the screw is under tension, the lifting distance, L, does not have any effect on the calculations. Thus, we can consider it as $L = 0$. In the left side of Fig. 6.8,

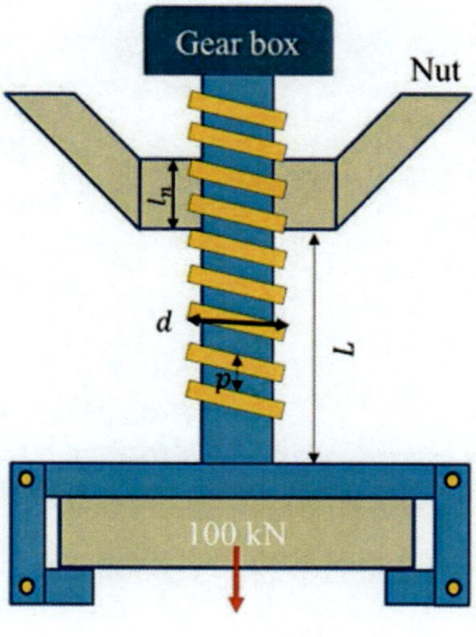

Fig. 6.9 Schematic of a power screw used for lifting a burden (by pulling it upwards)

we find $F = 100$ kN (on the vertical axis), go right to cross the curve of "*Border line*" and then go down to reach the diameter, $d = 59$ mm.

2. Since L does not affect the calculations, we do not need to find the corresponding diameter for the double-thread screw using Fig. 6.6. Therefore, the screw should have this characteritics: $60 \times 18P9$ and $l_n = 2.5d = 150$ mm.

6.3 Equations

$$F_1 = \frac{\sigma_p}{\sigma_t}$$
$$\sigma_p = \begin{cases} 0.2R_m & \text{(for one - way loading)} \\ 0.13R_m & \text{(for two - way loading)} \end{cases}$$
$$\sigma_t = \begin{cases} 5.055d^{-2.2} & \text{(for signle - thread)} \\ 6.608d^{-2.23} & \text{(for double - thread)} \\ 8.5022d^{-2.267} & \text{(for triple - thread)} \end{cases} \tag{6.1}$$

$$F_2 = \frac{0.5\pi l_n d_2}{p_p}$$
$$l_n = 2.5d$$
$$d_2 = 0.948d - 1.086$$
$$p_p = 5,\ 8,\ 10 \text{ and } 5\,\text{MPa (for cast iron, steel, bronze and plastic, respectively)} \tag{6.2}$$

$$\text{Euler:}\ F_3 = \frac{\frac{210000\pi^2}{\lambda^2}}{S\sigma_t}$$
$$\lambda \approx \frac{2.8L}{d_3}$$
$$S = 6 \tag{6.3}$$

$$\text{Johnson:}\ F_3 \approx \frac{0.8R_e}{S\sigma_t}$$
$$S = 35 \tag{6.4}$$

$$T = \begin{cases} F(0.088d + 0.278)/1000 & \text{(for signle - thread)} \\ F(0.1053d + 0.65)/1000 & \text{(for double - thread)} \\ F(0.1226d + 1.0358)/1000 & \text{(for triple - thread)} \end{cases} \tag{6.5}$$

References

Budynas RG, Nisbett JK (2015) Shigley's mechanical engineering design. McGraw-Hill, New York

Collins J, Busby H, Staab G (2009) Mechanical design of machine elements and machines: a failure prevention perspective, 2nd edn. Wiley, New York

Decker K-H, Kabus K (2014) Decker Maschinenelemente: Funktion, Gestaltung und Berechnung. Carl Hanser Verlag, München (in German)

Juvinall R, Marshek K (2011) Fundamentals of machine component design, 5th edn. Wiley, New York

Mott RL, Vavrek EM, Wang J (2018) Machine elements in mechanical design, 6th edn. Pearson Education, New York

Chapter 7
Shafts

Nomenclature

d	Shaft diameter, mm
R_m	Ultimate tensile strength, MPa
HB	Brinell Hardness
S	Safety factor
k	The factor of size, stress concentration, roughness, and type of load on shaft
T	Torque, N·m
M	Bending moment, N·m
β	Stress concentration factor
M_v	Equivalent moment, N·m
σ_b	Bending fatigue strength, MPa

7.1 Introduction

A shaft is an important machine element that transmits rotational power and/or motion. Components such as gears, pulleys, flywheels, cranks, sprockets and couplings, are mounted on or attached to the shaft in order to perform the transmissions.

Shafts are mostly cylindrical (solid or hollow). Shafts are usually rotating elements and are attached to a fixed frame or housing by support of bearings (called transmission shafts). These shafts are usually stepped and bear both bending and torsional moments. There are rotating shafts that only carry torsional moments (called spindles). In some applications, shafts are stationary and support rotating members including *idler* gears, pulleys and wheels, through bearings (called axles). Axles

M. Yaghoubi and H. Tavakoli, *Mechanical Design of Machine Elements by Graphical Methods*, Materials Forming, Machining and Tribology,
https://doi.org/10.1007/978-3-031-04329-1_7

only take bending moments (Budynas and Nisbett 2015; Jiang 2019; Juvinall and Marshek 2011; Mott et al. 2018).

7.2 Design Methods

Shaft design involves specifying suitable dimensions for the shaft that ensure its satisfactory and safe operation. The design process must consider the design of the mounted elements simultaneously. Because the elements are interdependent. Therefore, as a first step in the process of designing the shaft system, a rough estimate of the shaft is done. Then, after designing the mounted elements, a precise design of the shaft can be performed. The following subsections present the methods proposed in this book for the shaft design.

7.2.1 Rough Estimation (M.Y Method)

In this book, we propose a novel graphical method for estimating the shaft diameter (the related equations can be found in Sect. 7.3 (Eq. 7.1)). Figure 7.1 shows a diagram representing the diameter as a function of the torque for shafts under different loading conditions. The following examples illustrate the use of the method.

Example 7.1 Consider a power transmission system as depicted in Fig. 7.2. The power generated by an engine transmits through a U joint (Cardan joint) to a gearbox including two shafts. Estimate the minimum diameters, d_1, d_2, d_3 and d_4.

Solution

1. We apply a safety factor of 1.5 to the torques:

$$T_1 = 1.5 \times 500 = 750\ \mathrm{N \cdot m}$$
$$T_2 = 1.5 \times 1000 = 1500\ \mathrm{N \cdot m}$$

2. d_1: since it pertains to a U-joint () and the shaft is under pure torsion (T_1 = 750 N·m), we use the curve No. 6 of Fig. 7.1. As shown, we draw a T = 750 N·m vertical line crossing the curve No. 6 and then draw a horizontal line from the point of intersection to obtain the corresponding diameter on the vertical axis that is 46 mm.

$$d_1 = 46\ \mathrm{mm}$$

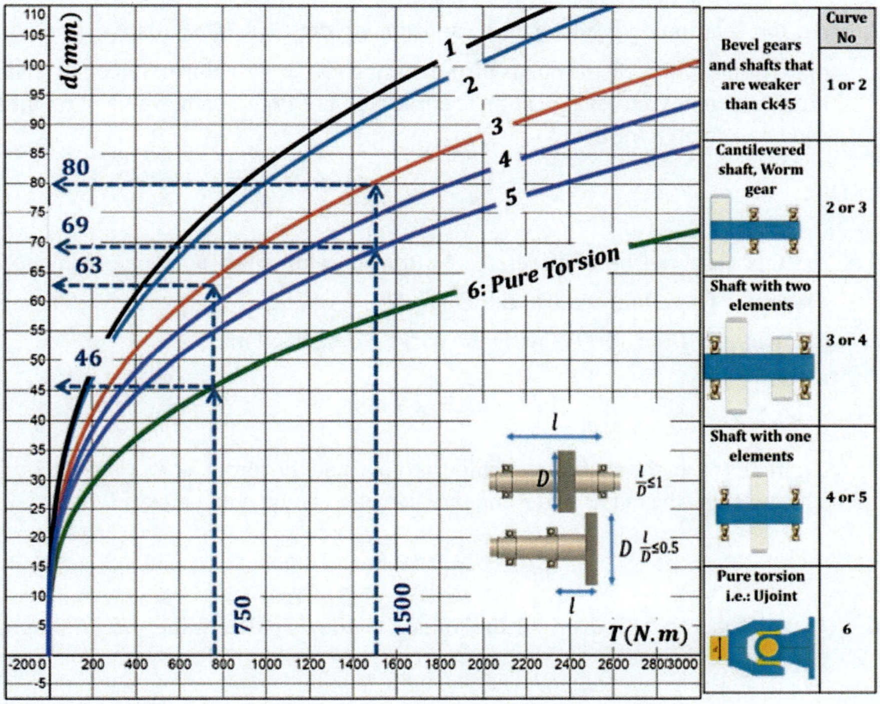

Fig. 7.1 The shaft diameter as a function of the torque for graphical estimates of the diameter

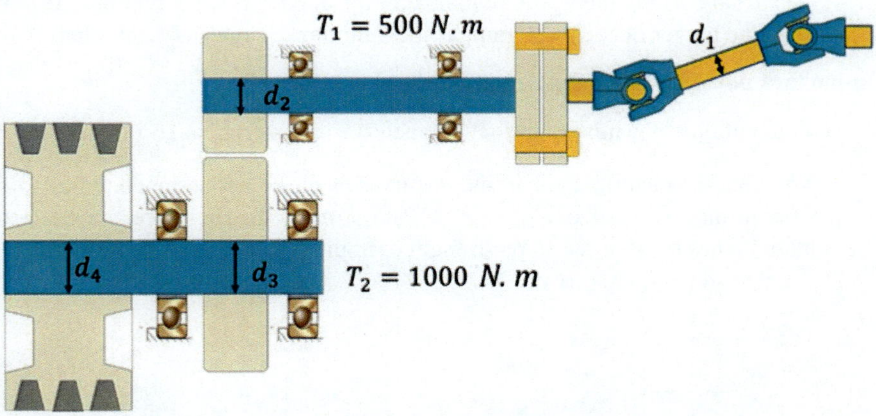

Fig. 7.2 The power transmission system of Example 7.1

3. d_2: the rotational speed for this cantilevered shaft () is the same as the U-joint. Thus, the torque is also the same ($T_1 = 750$ N·m). Since the shaft arrangement is cantilever, using the curve No. 3 of Fig. 7.1, the value of 63 mm can be estimated for d_2.

$$d_2 = 63 \text{ mm}$$

4. d_3: this case can be considered as a transmission shaft with one element (). Therefore, we use the curve No. 5 of Fig. 7.1. The corresponding diameter for $T_2 = 1500$ N·m on the curve No. 5, is 69 mm.

$$d_3 = 69 \text{ mm}$$

5. d_4: in this case, the shaft is a cantilevered one and the torque is $T_2 = 1500$ N·m. Thus, we use the curve No. 3 and the diameter can be obtained as 80 mm.

$$d_4 = 80 \text{ mm}$$

6. At the end, we round up all the values to roughly determine the minimum diameter of the shafts:

$$d_1 = 50 \text{ mm}, \ d_2 = 65 \text{ mm}, \ d_3 = 70 \text{ mm, and } d_4 = 80 \text{ mm}$$

7. Since d_3 and d_4 are the values obtained for the diameter of a common shaft, we select the bigger one, $d_4 = 80$ mm, as the minimum diameter for this shaft.

Important notes for using M.Y method and Fig. 7.1:

i. Our intention of the minimum shaft diameter, is the size, d_A, as shown in Fig. 7.3

ii. Note that the material of shaft in the curves of Fig. 7.1 is assumed to be Ck45 with ultimate strength of $R_m = 600$ MPa (normalized). Therefore, the selected material has to be of the same strength or higher.

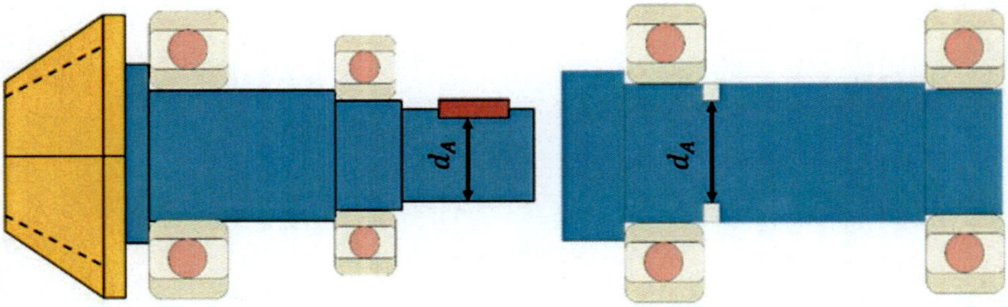

Fig. 7.3 An illustration for the minimum shaft diameter

iii. Some designers usually use hollow shafts for the U-joints. In this respect, the maximum size of the inner diameter (d_{in}) should be $0.6d$ ($d_{in} \leq 0.6d$) (Orlov 1988)

iv. The curves of Fig. 7.1 are valid under condition of $\frac{l}{D} \leq 1$ for the simply supported shafts and $\frac{l}{D} \leq 0.5$ for the cantilevered shafts. However, for other conditions, instead of using the relevant curve to the given shaft, use the upper curve. For example for shaft d_4 in Example 7.1, instead of the curve No. 3, use the upper curve, i.e. No. 2.

v. Figures 7.4 and 7.5 show several shaft arrangements that can guide you to find the equivalent torque (T) and the best relevant curve for using Fig. 7.1.

Note In machines, you can rarely find shafts, which are under pure torsion. The examples of their application are propeller shafts, shafts connected to pumps and centrifugal fans, etc. (Wittel et al. 2013). Therefore, it is recommended to always consider shafts under bending and torsion.

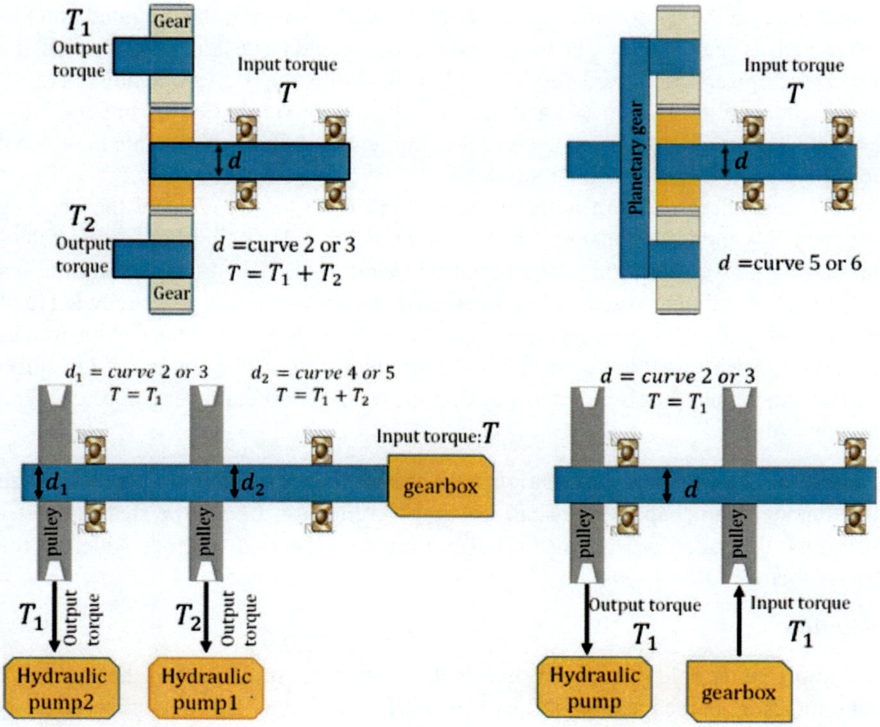

Fig. 7.4 Different transmission systems (V-belt, gearbox and planetary gear) and their equivalent torque (T) and curve number

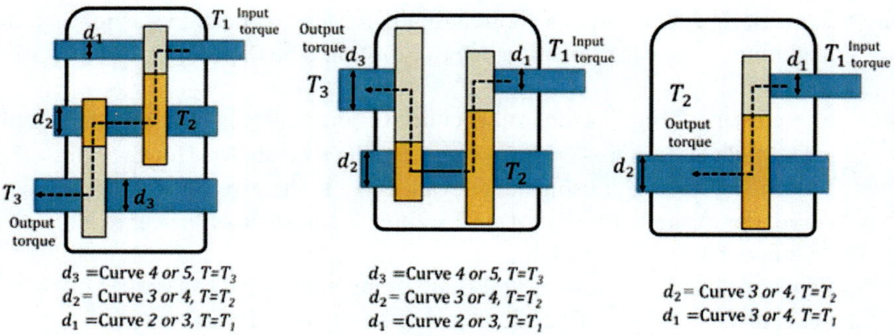

Fig. 7.5 Gearboxes with two and three stage reducers and their equivalent torque (T) and curve number

7.2.2 Precise Design (Checking)

Subsequent to the approximate estimate of the shaft diameter, the mounted components such as gears, pulleys, bearings and the like should be designed; afterwards, the precise design can be carried out. It includes calculating the equivalent moment (M_v), stress-concentration factor (β) and the shaft diameter (d) (see Eq. 7.2 in Sect. 7.3).

Note The precise design is useful for complicated shaft arrangements in which a lot of elements are mounted on the shaft.

The equivalent moment is the combination of the torque (T) and the bending moment (M) and is computed by the equation: $M_v \approx \sqrt{M^2 + T^2}$. For precise calculation of M_v, we refer you to the literature (Mott et al. 2018); the calculation is a complicated and time-consuming process and you should have a good knowledge of *statics*. Instead, we propose a graphical method shown in Fig. 7.6 that can be used as a straightforward method for simple shaft arrangements. After obtaining the equivalent moment and the stress-concentration factor, the shaft diameter can be checked and designed.

Example 7.2 After rough estimation of the shaft diameter in Example 7.1, we finalize the diameter and shape of the shaft as shown in Fig. 7.7. Check the shaft diameter using the described precise method. The torque on the shaft with considering the safety factor is $T = 1500$ N·m.

Solution

The most vulnerable part of the shaft is the sharp fillet shoulder and the groove of retaining ring, in the location of "*Bearing 1*". Thus, we only check the diameters, $d_2 = 85$ mm and $d_3 = 80$ mm.

First, we should find the equivalent moment (M_v) in the location of "*Bearing 1*".

1. We calculate $\frac{L}{D}$ for Fig. 7.7:

$$\frac{L}{D} = \frac{150 \text{ mm}}{300 \text{ mm}} = 0.5$$

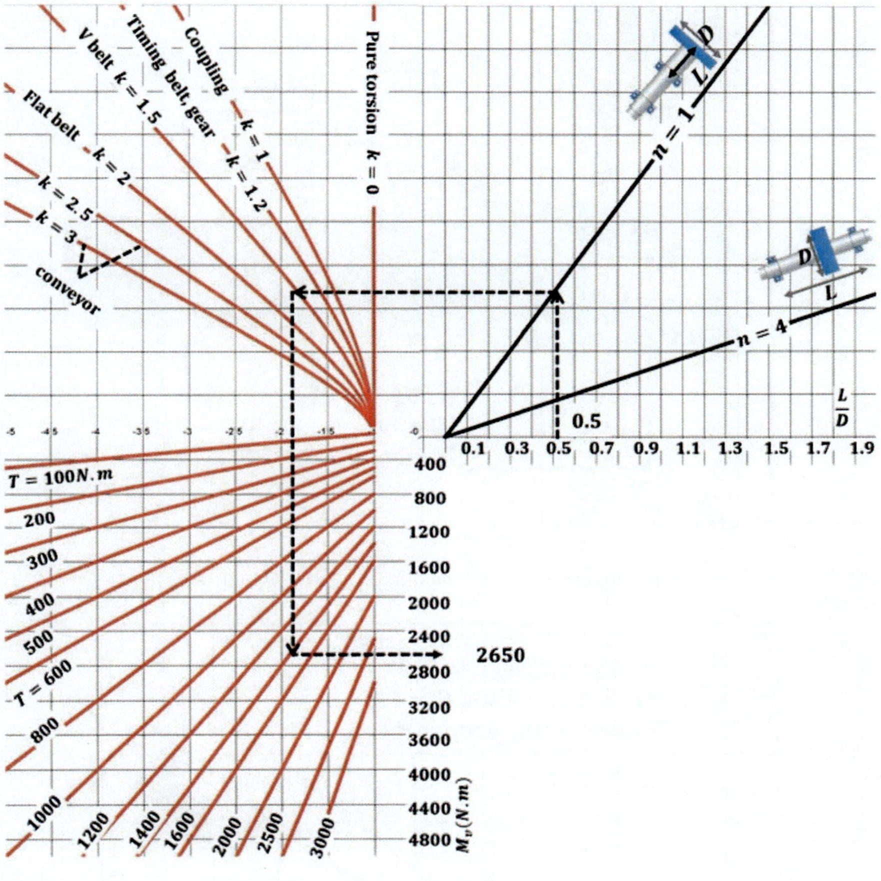

Fig. 7.6 Graphs for estimate of the equivalent moment

2. From the right horizontal axis of Fig. 7.6, we choose $\frac{L}{D} = 0.5$ and go up to intersect the curve related to the cantilevered shaft () (because here the shaft arrangement is cantilevered).
3. Then, we go left to intersect the curve of *V-belt pulley*.
4. We go down from the intersected point, to reach the approximate curve of $T = 1500$ N·m, and then go right to find the equivalent moment on the lower vertical axis:

   $M_v \approx 2650$ N.m .
5. Now, using Fig. 7.8, the stress-concentration factor (β) can be found.

   For the sharp fillet shoulder: $\beta = 2.5$.

   For the retaining ring groove, $\beta = 3$.

   Note R_m (MPa) is the ultimate strength of shaft. If the value of R_m is not given, you can calculate it using $R_m = (3.2 - 3.4)HB$ (see Eq. 7.2 in Sect. 7.3).

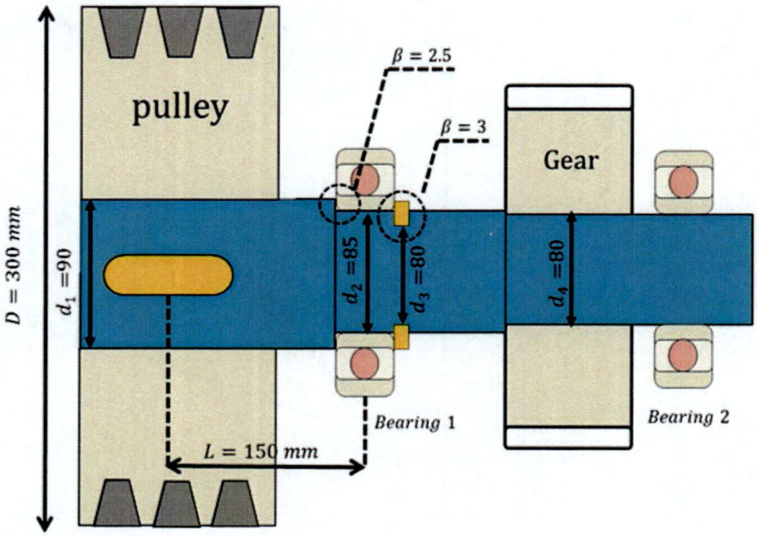

Fig. 7.7 The final specified dimensions of the shaft in Examples 7.1 and 7.2

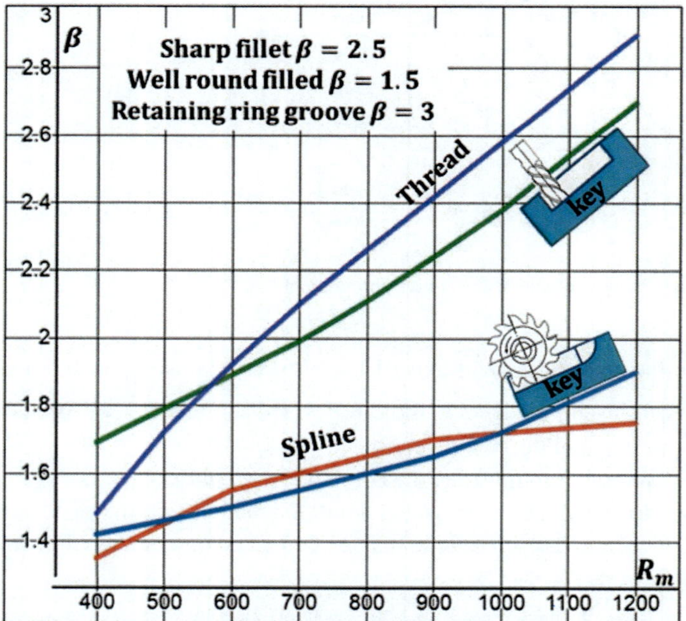

Material	St37	E295	C35,	E335	Ck45	20Cr4	36NiCr6,	40NiCr6,	42Cr4	42CrMo4	20MnCr5
R_m	340	500	600		710	780	980			1100	1200
			Heat treated								Case hardened

Fig. 7.8 A graph for estimating the stress-concentration factor (β)

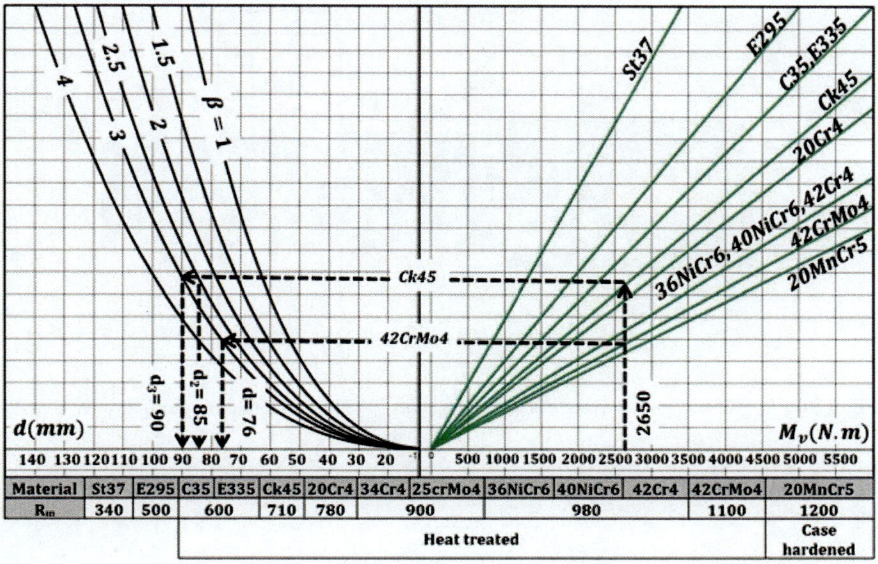

Fig. 7.9 Graphs for checking of the shaft diameter

6. To obtain the shaft diameters, in Fig. 7.9, we find $M_v = 2650$ N·m, then go up vertically to the corresponding point on the curve of *Ck45* (selected material for the shaft).
7. From this point we draw a horizontal line so that it crosses the curves of $\beta = 2.5$ and 3.
8. At the end, we go down to obtain the shaft diameters, $d_2 = 85$ mm and $d_3 = 90$ mm.

The obtained values must be less than or equal to the specified values in Fig. 7.7. The value of d_2 in this example ($d_2 = 85$ mm) is equal to the specified value for d_2 (85 mm) in Fig. 7.7. Therefore, the diameter, $d_2 = 85$ mm is suitable.

However, for d_3 which is related to the retaining ring groove, the obtained value is higher than the specified one in Fig. 7.7. This means that the defined diameter $d_3 = 80$ mm in Fig. 7.7 is not sufficient and must be at least 90 mm. This is important, because the stress-concentration near the bearings caused by high bending moments can be unsafe. If we change d_3 from 80 to 90 mm, we have to redesign the bearing and other components too. To avoid this, we can apply one of the following ways:

- Increasing the distance between the groove and the bearing or removing the retaining ring groove and replacing it by a spacer (Fig. 7.10).
- Changing the material of shaft to a stronger one such as 42CrMo4 that decrease the values obtained for the diameters (Fig. 7.9). As a result, with 42CrMo4, shaft diameter is d = 76 mm that is less than specified value $d_3 = 80$ mm in Fig. 7.7.

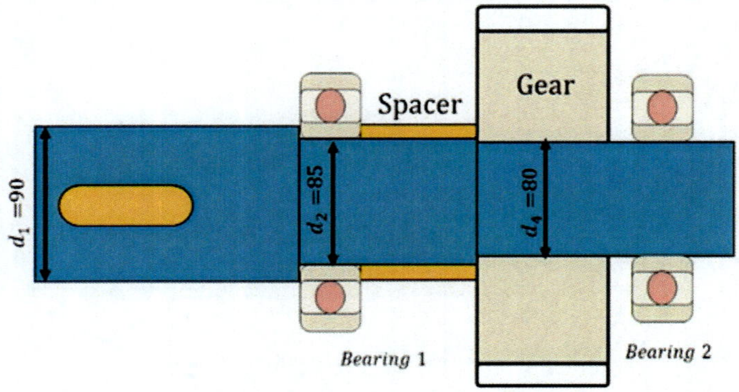

Fig. 7.10 An illustration for using a spacer on a shaft instead of the retaining ring groove

Fig. 7.11 An illustration for the retaining rings

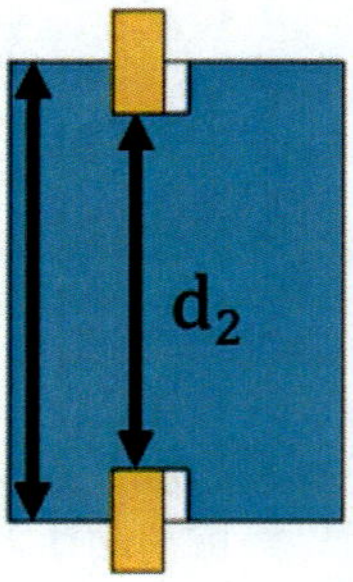

Note In Example 7.1 for determining d_4, between the curves No. 2 and No. 3 for the cantilever shaft, we used curve No. 3. If we use the curve No. 2, the minimum diameter of d_4 would be 90 mm that is suitable too.

Note In retaining rings: $d \approx 1.05d_2$ (Fig. 7.11). This means that after calculating the minimum shaft diameter (d), if it is necessary to add a retaining ring to the shaft, it is better to multiply the minimum diameter by a factor of 1.05.

Sometimes designers use polygonal shafts for power transmission. For the design of these types of shafts, you need to calculate them under crushing stress. However, in this chapter, you can easily estimate the shaft diameter (d) from Fig. 7.1. To find the inscribed circle diameter (d) of the triangular and rectangular profiles, use the curve No. 5 or No. 4, and for the hexagonal profile, use the curve No. 3.

After finding the inscribed circle diameter (d), you can find the circumscribed circle diameter (D_c) from Fig. 7.12.

Note If the shaft is connected to a hub and the hub moves along the shaft (sliding when loaded), multiply the obtained shaft diameter (d) by 1.6 and the connection length (l) of the hub and the shaft should be $l \geq d$.

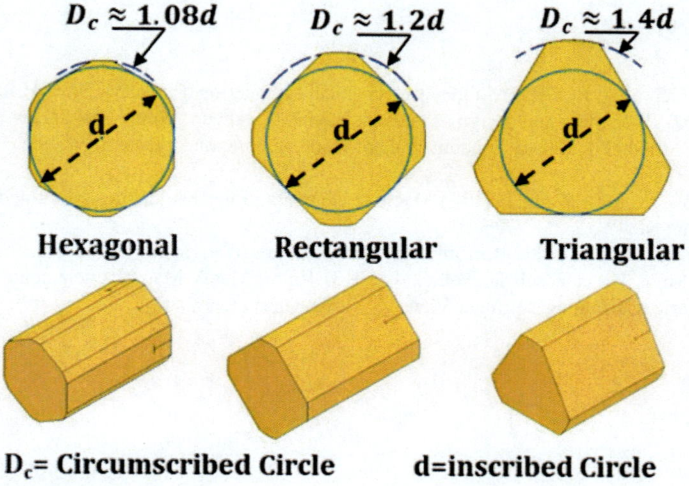

Fig. 7.12 Polygonal shafts

7.3 Equations

M.Y method (rough estimation)

$$\begin{aligned} &\mathrm{d} = m\sqrt[3]{T} \\ &m = \begin{cases} 5 & (\text{ for curve No.6}) \\ 6 & (\text{for curve No.5}) \\ 6.5 & (\text{ for curve No.4}) \\ 7 & (\text{for curve No.3}) \\ 8 & (\text{for curve No.2}) \\ 8.3 & (\text{for curve No.1}) \end{cases} \end{aligned} \tag{7.1}$$

(on Fig. 7.1)

Precise design (checking)

$$\begin{aligned} &S \times 1000M_v = 0.1d^3\sigma_b k \\ &M_v \approx \sqrt[2]{M^2 + T^2} \\ &\sigma_b = 0.5R_m \\ &R_m = (3.2 - 3.4)HB \\ &S = 2 \\ &k = \frac{1 - 0.26\log\frac{d}{16}}{\frac{\beta}{1-0.2\frac{\log\frac{d}{7.5}}{\log 20}} + \frac{1}{0.85} - 1} \approx \left(1.6\beta^{-0.8} - 0.15\right)d^{-0.2} \end{aligned} \tag{7.2}$$

References

Budynas RG, Nisbett JK (2015) Shigley's mechanical engineering design. McGraw-Hill, New York
Jiang W (2019) Analysis and design of machine elements, 1st edn. Wiley, New Jersey
Juvinall R, Marshek K (2011) Fundamentals of machine component design, 5th edn. Wiley, New York
Mott RL, Vavrek EM, Wang J (2018) Machine elements in mechanical design, 6th edn. Pearson Education, New York
Orlov P (1988) Fundamentals of machine design. Mechanical engineering, Moscow (in Russian)
Wittel H, Muhs D, Jannasch D, Voßiek J (2013) Roloff/Matek Maschinenelemente Normung, Berechnung, Gestaltung. Springer Vieweg, Wiesbaden (in German)

Chapter 8
Rectangular Keys

Nomenclature

b	The width of key, mm
d_{sh}	Shaft diameter, mm
h	The height of key, mm
L	The effective (or working) length of key, mm
L_s	Standard length of key, mm
P_0	Allowable bearing stress, MPa
R_e	Yield strength, MPa
T	Torque, Nm
t_1	The effective (or working) height of key, mm

8.1 Introduction

A key is a detachable component used to connect a shaft and the hub of a power-transmitting element, which enables transmission of torque and power between them. Rectangular keys, also called flat keys, are usually used for large shafts and are also recommended for small shafts where the shorter key height can be tolerated (Jiang 2019; Mott et al. 2018).

M. Yaghoubi and H. Tavakoli, *Mechanical Design of Machine Elements by Graphical Methods*, Materials Forming, Machining and Tribology,
https://doi.org/10.1007/978-3-031-04329-1_8

8.2 Design Methods

Designing the key includes determining its dimensions (height (h), width (b) and length (L)) according to the transmitted torque (T) and the shaft diameter (d_{sh}). Equation 8.1 (see Sect. 8.4) is used for this purpose (also see Fig. 8.1).

In this chapter, three methods are presented for designing rectangular keys.

8.2.1 Method I

This method can be applied to the cases in which the shaft has been designed beforehand. Instead of using the equation, we created graphs for this purpose. The following example shows the use of the method.

Example 8.1 A power is supposed to be transmitted to a fan through a belt and pulleys (Fig. 8.2). Find the size of key (the height and width). The effective length of key is equal to the shaft diameter, $L = d_4 = 80$ mm.

Solution

Figure 8.3 can be used for determining the size of the key. The key sizes that are shown below the right horizontal axis, are based on DIN standard.

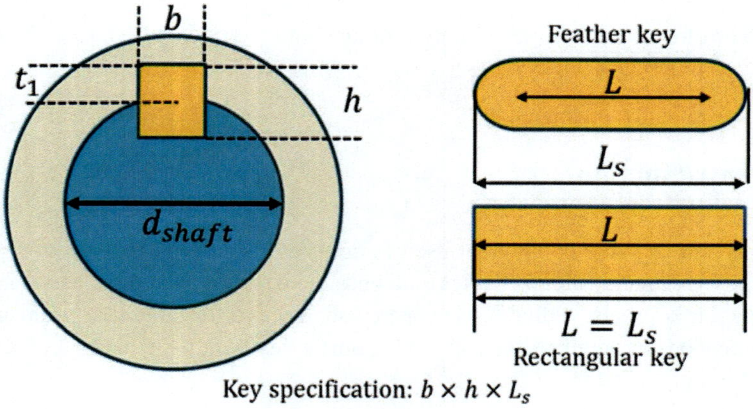

Fig. 8.1 Specifications of a rectangular key

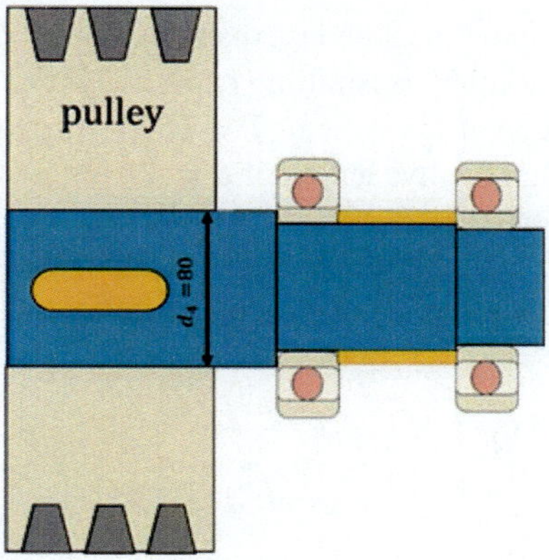

Fig. 8.2 Schematic of a pulley connected to a shaft by a rectangular key (Example 8.1)

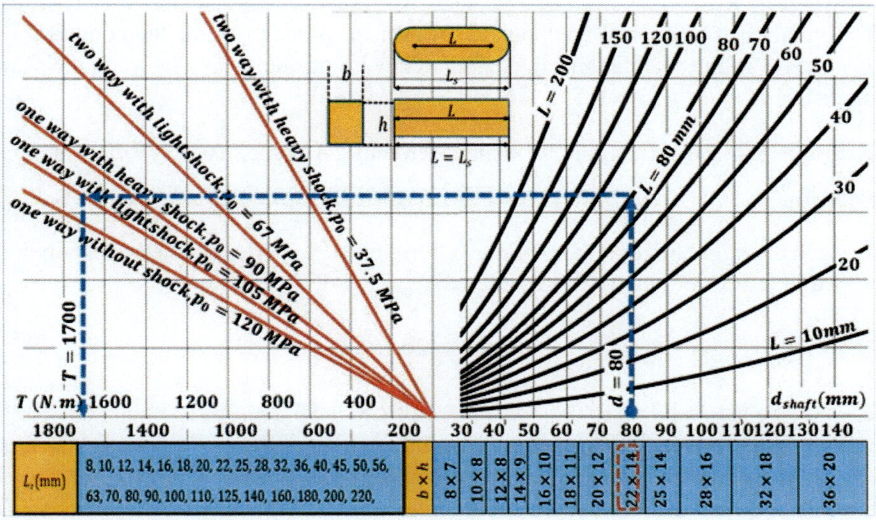

Fig. 8.3 Graphs for designing rectangular keys (Method I)

1. On the right horizontal axis of Fig. 8.3, we choose $d_{sh} = 80$ mm. As shown in Fig. 8.3 (below the horizontal axis related to d_{shaft}), the corresponding size of key for this diameter is $b \times h = 22 \times 14$ mm.
2. Then we go up to intersect the curve of $L = 80$ mm, from there go left to reach the curve of "*One way, light shock*". We use this curve because fans work one-way under light shock when they start. From the intersection point, we go down to obtain the torque transmission capacity of the key:

$$T = 1700 \text{ N.m.}$$

Note: The length of key must be less than or equal to $1.3d_{sh}$ ($L \leq 1.3d_{sh}$).

3. Since the obtained torque-transmitting capacity ($T = 1700$ N.m) is greater that the transmitted torque of the system ($T = 1500$ N.m), the key with the size $b \times h = 22 \times 14$ and the effective length of $L = 80$ mm is suitable for this example.
4. The specifications of a key is $b \times h \times L_s$. For rectangular keys $L_s = L$ and for feather keys $L_s \geq L$. If you are intended to choose a feather key, then use one or two size larger than the standard length (L_s) of rectangular key from Table 8.1. Thus:

$$\text{For rectangular key: } b \times h \times L_s \xrightarrow{L=L_s=80} 22 \times 14 \times 80 \text{ mm}$$
$$\text{For feather key, } b \times h \times L_s \xrightarrow{L_s=\text{one or two size larger than } L=80} 22 \times 14 \times 100 \text{ mm.}$$

8.2.2 Method II

This method is similar to Method I. However, in order to fulfill the condition of $L \leq 1.3d_{sh}$ in the calculations, the length of key is considered as a factor of the shaft diameter: $L = nd_{sh}$ ($n = 0.5$, 1, 1.3, or 1.5).

Example 8.2 Solve Example 8.1 using Method II. Assume that $d_{sh} = 70$ mm.

Solution

1. On the right horizontal axis of Fig. 8.4, we find $d_{sh} = 70$ mm. The corresponding dimensions of key for this diameter is

$$b \times h = 20 \times 12 \text{ mm}$$

2. We go vertically up to cross the curve of $L = 1.3d_{shaft}$.
   Note: It is recommended to choose the length of the rectangular key as $L = 1.3d_{shaft}$.
3. Then go left to intersect the curve of "*One way, light shock*". After that, going down to the horizontal axis yields the torque transmission capacity of the key:

$$T = 1600 \text{ N.m.}$$

4. Since the obtained torque is greater than the applied torque ($T = 1500$ N.m), the selected dimensions for the key is appropriate.
5. The length of key is:

$$L = 1.3d_{sh} = 1.3 \times 70 = 91 \text{ mm}$$

   According to the standard lengths for keys (Table 8.1), we round this value up to 100 mm.
6. Therefore, the dimensions of the key for this example using Method II is:

Table 8.1 The standard lengths (L_s) for rectangular keys

L_s (mm)	8–10–12–14–16–18–20–22–25–28–32–36–40–45–50–56–63–70–80–90–100–110–125–140–160–180–200–220–250–280–315–355–400

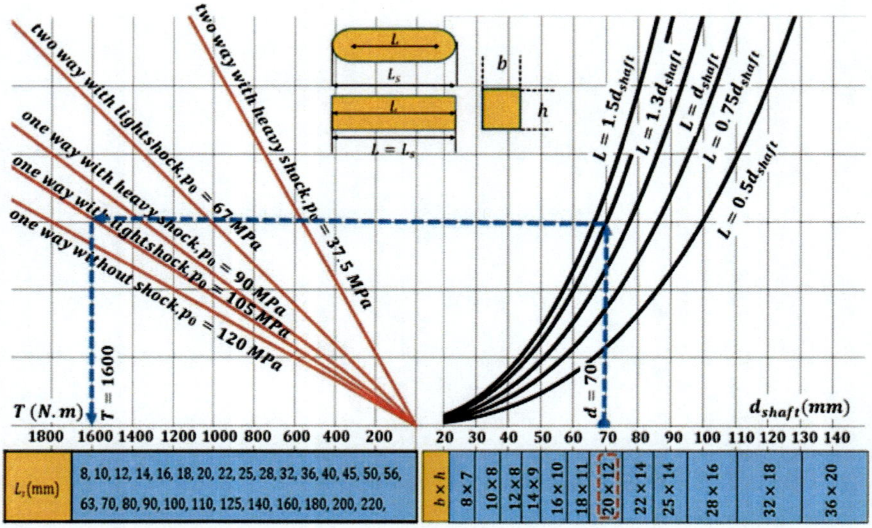

Fig. 8.4 Graphs for designing rectangular keys (Method II)

For rectangular key: $b \times h \times L_s \xrightarrow{L_s=100} 20 \times 12 \times 110$ mm

For rectangular key: $b \times h \times L_s \xrightarrow{L_s=100} 20 \times 12 \times 110$ mm

If in step 2 we choose the curve of $L = d_{shaft}$, $T \approx 1150$ N.m will be obtained which is lower than the applied torque ($T = 1500$ N.m). Therefore, $L = 1.3d_{sh}$ was an appropriate selection.

8.2.3 Method III (M.Y Method)

This method, named as M.Y method, is proposed in this book. The assumption of the method is that the shaft diameter is unspecified. Therefore, we determine the shaft diameter and the size of key simultaneously. The graphs of Fig. 8.5 can be employed for this method. The procedure of applying the graphs is illustrated in the following example.

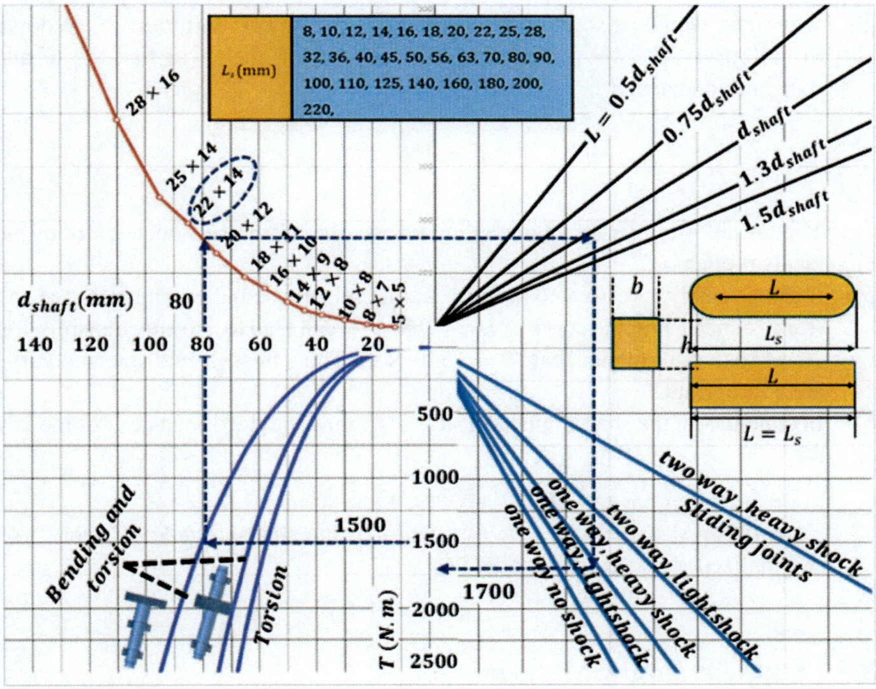

Fig. 8.5 Graphs for designing rectangular keys (M.Y method)

Example 8.3 Solve Example 8.1 using M.Y method. Assume that the shaft diameter is unknown.

Solution

1. On the lower vertical axis of Fig. 8.5, we choose $T = 1500$ N.m, and go left to reach the curve related to 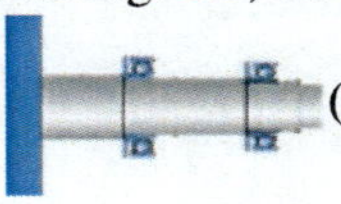 (because the shaft is under torsion and bending).
2. Then, we go up to find the shaft diameter on the left horizontal axis:

$$d_{sh} = 80 \text{ mm}$$

3. We keep going up to cross the curve in the second quadrant (the upper left corner) of Fig. 8.5. On this curve, we can find the size of key. Since the intersection point, in this example, is between $b \times h = 20 \times 12$ and $b \times h = 22 \times 14$, we select the greater one.

4. After that, we go right to intersect $L = d_{shaft}$ (we select it arbitrarily), go down to the curve of "*One way, light shock*", and then go left to find the torque transmission capacity:

$$T = 1700\ \text{N.m}$$

The torque is bigger than the applied torque, therefore, the selected size of the key is proper.

5. It can be easily investigated in Fig. 8.5, that selecting smaller length for the key (for instance using the curve of $L = 0.75d_{shaft}$) can lead to a torque transmission capacity that is smaller than $T = 1500$ N.m. Thus, the selected size and curve are appropriate.
6. In conclusion, the shaft diameter is $d_{sh} = 80$ mm and the dimensions of the key are:

For rectangular key: $b \times h \times L = 22 \times 14 \times 80$ mm

Note: in M.Y method, it is assumed that the shaft is made of steel Ck45 (normalized) or a stronger material.

8.3 Supplementary Material

Supplementary Table 8.1 presents allowable bearing stress values from different references (Orlov 1988; Zhukov and Gurevich 2014 ; Mott et al. 2018).

8.4 Equations

$$T = \frac{p_0 L \times 0.5 d_{sh} t_l}{1000}$$

$$t_1 \approx 0.45h$$

Supplementary Table 8.1 Allowable bearing stress (P_0) based on different references

Reference	Hardness	Permissible bearing stress (MPa)		
		Constant torque	Pulsating torque	Alternating torque
Orlov (1988)	≥ 30 HRC	300–400	200–300	100–200
	< 30 HRC	150–200	100–150	50–100
For sliding joint divide the values above by 2 or 3				
Zhukov and Gurevich (2014)	All hardnesses	$P_0 = 0.6R_e$	$P_0 = 0.4R_e$	$P_0 = 0.3R_e$
Mott et al. (2018)	All hardnesses	$P_0 = \frac{R_e}{3}$		

$$
\begin{aligned}
h &\approx 0.1d_{sh} + 3.594 \\
0.5d_{sh}t_l &\approx 0.022d_{sh}^2 + 0.7474d_{sh} + 0.635
\end{aligned}
\tag{8.1}
$$

References

Autodesk Inc (2021) Engineer's handbook. Retrieved from Autodesk Inventor support and learning center: https://knowledge.autodesk.com/

Böge A (2011) Handbuch Maschinenbau: Grundlagen und Anwendungen der Maschinenbau-Technik. Vieweg+Teubner, Wiesbaden (in German)

Decker K-H, Kabus K (2014) Decker Maschinenelemente Funktion, Gestaltung und Berechnung. Carl Hanser Verlag, München (in German)

Haberhauer H (2018) Maschinenelemente: Gestaltung, Berechnung, Anwendung. Springer Vieweg, Berlin, Heidelberg (in German)

Jiang W (2019) Analysis and design of machine elements, 1st edn. Wiley, New Jersey

Klebanov BM, Barlam DM, Nystrom FE (2007) Machine elements: Life and Design, 1st edn. CRC Press, Boca Raton

Mott RL, Vavrek EM, Wang J (2018) Machine elements in mechanical design, 6th edn. Pearson Education, New York

Orlov P (1988) Fundamentals of Machine Design. Mechanical engineering, Moscow (in Russian)

Schlecht B (2015) Maschinenelemente 1. Pearson Education, London (in German)

Wittel H, Muhs D, Jannasch D, Voßiek J (2013) Roloff/Matek Maschinenelemente Normung, Berechnung, Gestaltung. Springer Vieweg, Wiesbaden (in German)

Chapter 9
Splines

Nomenclature

D	Outer diameter, mm
d	Inner diameter, mm
h	The effective (or working) height of spline, mm
K	Coefficient of equivalent torque
L	The effective (or working) length of spline, mm
m	Module, mm
n	Number of teeth
P_0	Allowable bearing stress, MPa
T	Torque, N·m
T_e	Equivalent torque, N·m

9.1 Introduction

Splines are in fact a series of keys machined on outside of a shaft, with corresponding grooves machined on inside of hub of a power-transmitting element (such as gear, sheave, sprocket, pulley etc.). The same as keys, splines are used to provide a joint connection between the hub and the shaft for torque and power transmission. However, they are typically used in applications in which a high torque needs to be transmitted.

They may have either straight-sided or involute profiles (Fig. 9.1). According to the standards, the straight-sided splines are typically made with 4, 6, 10 or 16 teeth; and in terms of the transmitted torque, they are classified into three categories: light-, medium- and heavy-duty. According to SAE, light, medium and heavy duty splines

M. Yaghoubi and H. Tavakoli, *Mechanical Design of Machine Elements by Graphical Methods*, Materials Forming, Machining and Tribology,
https://doi.org/10.1007/978-3-031-04329-1_9

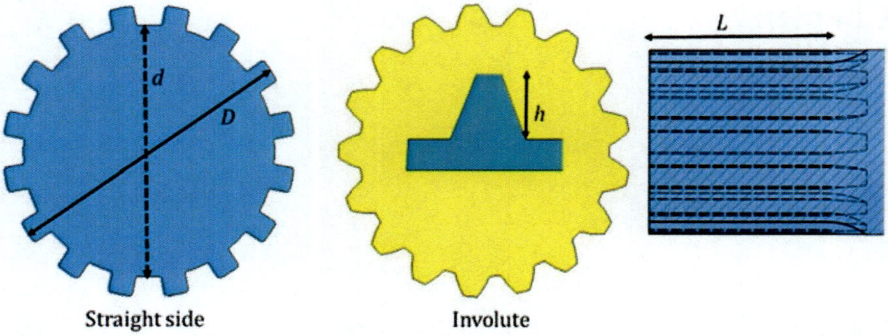

Fig. 9.1 Types of splines

are used for non-sliding, sliding without loading and sliding when loaded, respectively. While, involute splines are generally made with pressure angles of 30°, 37.5°, or 45°. Involute splines are the widely used and preferable type in modern machinery. Because, they are usually stronger, provide self-centering of the joined elements and can be machined with standard hobs used for cutting gear teeth (Budynas and Nisbett 2015; Jiang 2019; Juvinall and Marshek 2011; Mott et al. 2018).

9.2 Design Method

Crushing on the working surface is the potential failure mode for a spline (Orlov 1988). Therefore, the design is usually based on the bearing strength of spline (see Eq. 9.1 in Sect. 9.3).

Table 9.1 presents a list of allowable bearing stresses for different operating conditions and joining types of splines.

Instead of using Eq. 9.1 (see Sect. 9.3), we propose a graphical method for designing splines. For more details about the graphs and the way they are applied, refer to the following examples.

Example 9.1 A U-joint is supposed to connect output of a diesel engine to a pump and run it with a torque of 300 N·m. (Fig. 9.2). Determine the size and type of the spline of the U-joint. Consider the effective length of the spline equal to the shaft diameter ($L = D$).

Solution

1. First, we should find the equivalent torque according to the operating conditions. Since the U-joint is connected to a diesel engine, which works with vibration, a little vibration and sliding can occur between the engaged teeth. To avoid this, the spline is usually lubricated by grease. Therefore, with assuming that grease is regularly applied, to be on the safe side, we apply a factor of 7.7 to the torque.

Table 9.1 The allowable bearing stresses for splines (Autodesk Inc 2021; Orlov 1988)

		Allowable bearing stress (P_0, MPa)*					
		Autodesk handbook			Orlov		
Type of joining		Operating condition	Non-hardened sides	Hardened sides	Operating condition	Non-hardened sides	Hardened sides
Non-sliding		Adverse	35–50	40–70	Adverse	10–20	20–30
		Medium	60–100	100–140	Medium	20–30	30–40
		Favorable	80–120	120–200	Favorable	30–40	40–60
Sliding	When loaded	Adverse	–	3–10	Adverse	3–7	7–10
		Medium	–	5–15	Medium	7–10	10–13
		Favorable	–	10–20	Favorable	10–13	13–20
	Without loading	Adverse	15–20	20–35	Adverse	5–10	10–15
		Medium	20–30	30–60	Medium	10–15	15–20
		Favorable	25–40	40–70	Favorable	15–20	20–30

* Non-hardened: hardness is less than or equal to 350 HB; Hardened: hardness is greater than 40 HRC

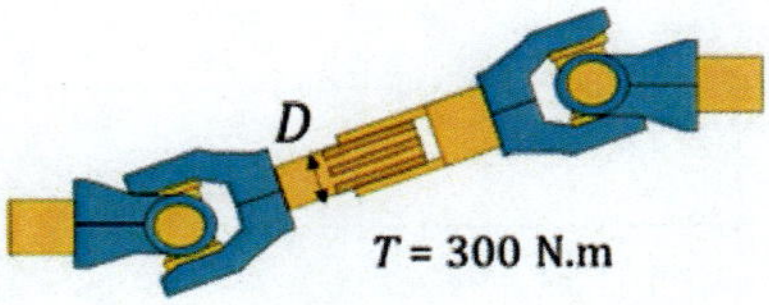

Fig. 9.2 A U-joint for transmitting torque (Example 9.1)

We obtained the factor from Table 9.2. For this example we applied method of "Zhukov":

$$T_e = K \times T = 7.7 \times 300 \approx 2300\ \mathrm{N \cdot m}$$

Note: SAE always uses $K = 7$ (Mott et al. 2018).

2. We choose $T_e = 2300$ N·m on the left horizontal axis of Fig. 9.3, go up to the curve of $L = D$, and then go right horizontally to cross all curves.
3. From each crossed curve, we go down to obtain the outer diameter of spline. As seen in Fig. 9.3, D = 50 mm achieved for *involute* spline and the *medium duty*, $n = 8$ and $D = 68$ mm obtained for *straight-sided* spline.
4. Now, we select the closest standard splines, which have outer diameters greater than or equal to those obtained in the previous step, according to Tables 9.3 and 9.4, for involute and straight-sided splines, respectively.

The following characteristics for the splines are resulted:
Involute spline: $D \times m = 50 \times 1.25$ or 50×2mm, and $L \geq D = 50$ mm.
Straight spline: $n \times d \times D = 8 \times 62 \times 72$ mm, and $L \geq D = 68$ mm.

Table 9.2 Coefficients for determining the equivalent torque of splines

Type of joining		Operating condition	Average coefficient of equivalent torque (K)*			
			Autodesk handbook		Zhukov	
			Non-hardened sides	Hardened sides	Non-hardened sides	Hardened sides
Sliding	When loaded	Adverse	–	7	–	10
		Medium	–	5	–	7.7
		Favorable	–	3.5	–	4.5
	Without loading	Adverse	3	2	4.2	2.7
		Medium	2	1.5	2.5	1.5
		Favorable	1.5	0.8	2	1.2
Non-sliding		Adverse	1.2	0.8	1.56	1.2
		Medium	0.6	0.5	0.83	0.5
		Favorable	0.5	0.4	0.6	0.4

* Non-hardened: hardness is less than or equal to 350 HB; Hardened: hardness is greater than 40 HRC

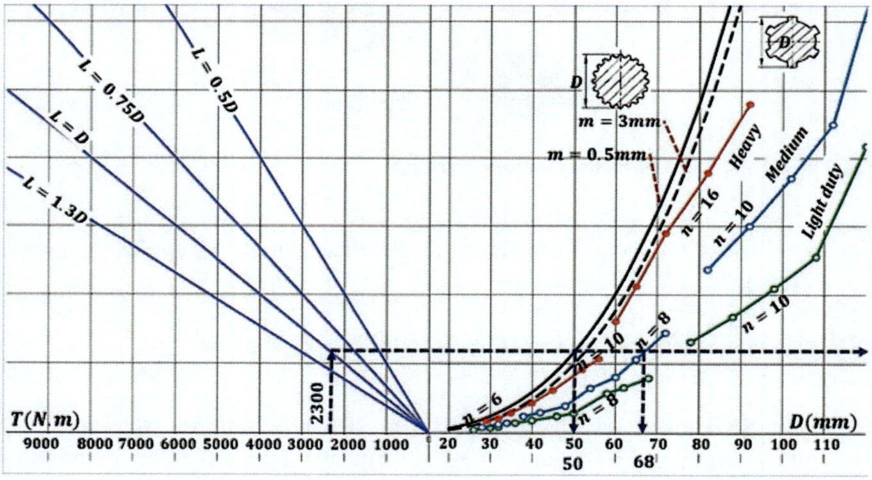

Fig. 9.3 Graphs for estimating outer diameter of spline based on the equivalent torque

Notes:

- For torques less than 1000 N·m, Fig. 9.4 can be employed.
- When the length of spline is known beforehand, and for instance we do not want to have it as a factor of outer diameter ($L = aD$), then Fig. 9.5 can be used.

Table 9.3 The standard values for dimensions of involute splines (Decker and Kabus 2014)

$m = 0.8$ mm*		$m = 1.25$ mm		$m = 2$ mm		$m = 3$ mm	
D	*n*	*D*	*n*	*D*	*n*	*D*	*n*
6	6	17	12	35	16	55	17
7	7	18	13	37	17	60	18
8	8	20	14	38	18	65	20
9	10	22	16	40	18	70	22
10	11	25	18	42	20	75	24
12	13	28	21	45	21	80	25
14	16	30	22	47	22	85	27
15	17	32	24	48	22	90	28
16	18	35	26	50	24	95	30
17	20	37	28	55	26	100	32
18	21	38	29	60	28	105	34
20	23	40	30	65	31	110	35
22	26	42	32	70	34	120	38
25	30	45	34	75	36	130	42
28	34	47	36	80	38	140	45
30	36	48	37			150	48
32	38	50	38				

* *m*, *D* and *n* are the module, outer diameter and number of teeth, respectively. All in mm

Table 9.4 The standard values for dimensions of straight-sided splines (Decker and Kabus 2014)

$n \times d \times D$*		
Light	Medium	Heavy
6 × 23 × 26	6 × 11 × 14	10 × 16 × 20
6 × 26 × 30	6 × 13 × 16	10 × 18 × 23
6 × 28 × 32	6 × 16 × 20	10 × 21 × 26
	6 × 18 × 22	10 × 23 × 29
8 × 32 × 36	6 × 21 × 25	10 × 26 × 32
8 × 36 × 40	6 × 23 × 28	10 × 28 × 35
8 × 42 × 46	6 × 26 × 32	10 × 32 × 40
8 × 46 × 50	6 × 28 × 34	10 × 36 × 45
8 × 52 × 58		10 × 42 × 52
8 × 56 × 62	8 × 32 × 38	10 × 46 × 56
8 × 62 × 68	8 × 36 × 42	

(continued)

Table 9.4 (continued)

$n \times d \times D$*		
Light	Medium	Heavy
	8 × 42 × 48	16 × 52 × 60
10 × 72 × 78	8 × 46 × 54	16 × 56 × 65
10 × 82 × 88	8 × 52 × 60	16 × 62 × 72
10 × 92 × 98	8 × 56 × 65	16 × 72 × 82
10 × 102 × 108	8 × 62 × 72	
10 × 112 × 120		20 × 82 × 92
	10 × 72 × 82	
	10 × 82 × 92	
	10 × 92 × 102	
	10 × 102 × 112	
	10 × 112 × 125	

* n, d and D are the number of teeth, inner diameter (mm), and outer diameter (mm), respectively

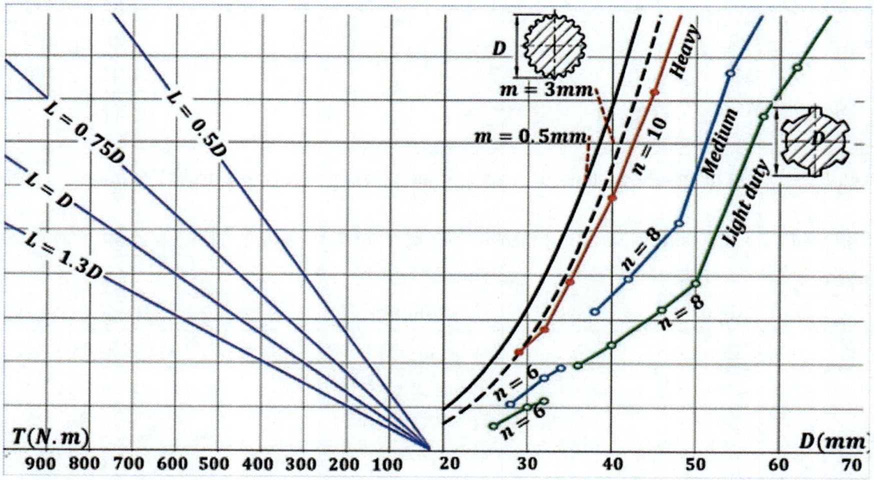

Fig. 9.4 Graphs for estimating outer diameter of spline based on the equivalent torques less than 1000 N·m

9.3 Equations

$$T = \frac{0.75 P_0 L h n D}{2000} = \begin{cases} \frac{0.75 P_0 L n (D-m)^2}{2000} & \text{(for involute splines)} \\ 0.75 P_0 L \frac{0.4(D-d)}{2000} n D & \text{(for straight splines)} \end{cases} \tag{9.1}$$

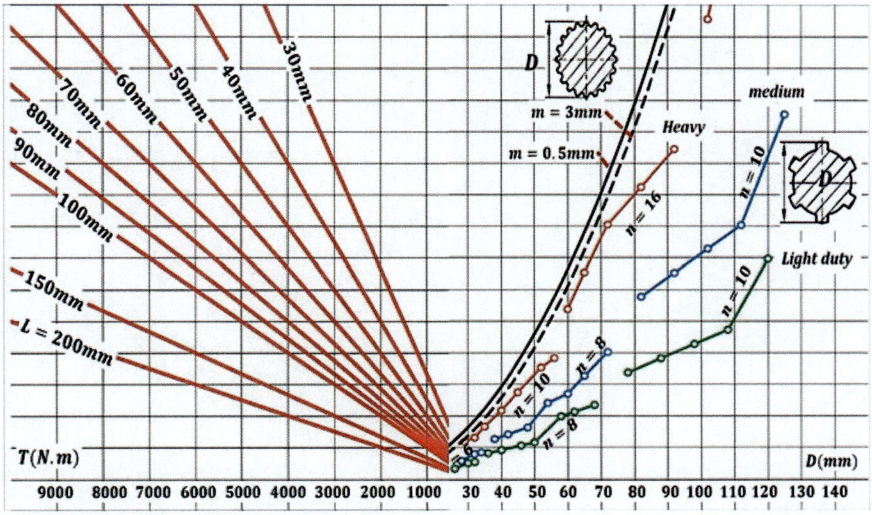

Fig. 9.5 Graphs for estimating outer diameter of spline based on the equivalent torques when the working length is known

References

Autodesk Inc. (2021) Engineer's handbook. Accessed from Autodesk Inventor support and learning center: https://knowledge.autodesk.com/

Budynas RG, Nisbett JK (2015) Shigley's mechanical engineering design. McGraw-Hill, New York

Decker K-H, Kabus K (2014) Decker Maschinenelemente: Funktion, Gestaltung und Berechnung. Carl Hanser Verlag, München (in German)

Jiang W (2019) Analysis and design of machine elements, 1st edn. Wiley, New Jersey

Juvinall R, Marshek K (2011) Fundamentals of machine component design, 5th edn. Wiley, New York

Mott RL, Vavrek EM, Wang J (2018) Machine elements in mechanical design, 6th edn. Pearson Education, New York

Orlov P (1988) Fundamentals of machine design. Mechanical engineering, Moscow (in Russian)

Chapter 10
Conical Joints

Nomenclature

C	Taper ratio ($C = 0.1$)
D	The larger diameter of frustum cone (or shaft diameter), mm
d	The smaller diameter of frustum cone, mm
d_{bolt}	Bolt diameter, mm
d_m	The median diameter of frustum cone, mm
F	The required force for pressing shaft and hub, N
F_{pre}	Allowable preload force of the bolt property class of 8.8 at $\mu = 0.12$, N
σ	Allowable contact stress, MPa
L	Connection length, mm
T	Torque, N·m
α	Cone angle, degrees ($\alpha = 5.724°$)
μ	Coefficient of friction ($\mu = 0.1$)
ρ	Angle of friction, degrees $\left(\rho = \tan^{-1}\mu = 6°\right)$
M	Allowable tightening torque of the bolt property class of 8.8 at $\mu = 0.12$, N·m

10.1 Introduction

Conical joint refers to a connection in which the shaft and hub are conical (in fact conical frustum) at the joint interface. The connected elements are pressed together by means of a bolt so that the torque and power can be transmitted by friction (Fig. 10.1). However, to avoid sliding caused by vibrations, keys are usually used for the connection too.

M. Yaghoubi and H. Tavakoli, *Mechanical Design of Machine Elements by Graphical Methods*, Materials Forming, Machining and Tribology,
https://doi.org/10.1007/978-3-031-04329-1_10

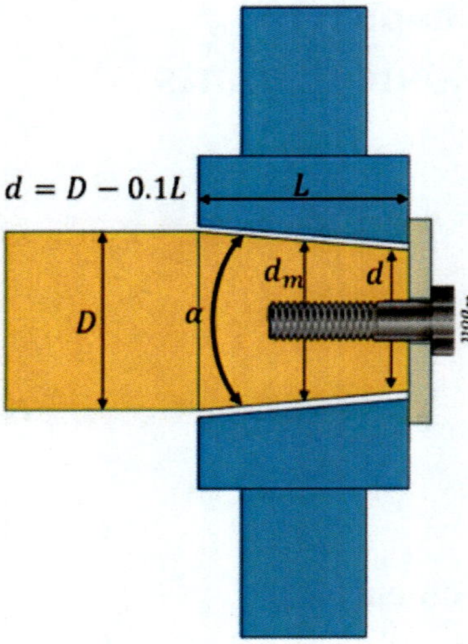

Fig. 10.1 Schematic of a conical joint

This chapter deals with estimating the maximum torque that can be transmitted solely by a conical joint. In addition, the suitable bolt size for fastening the hub and shaft is determined.

10.2 Design Methods

Equations 10.1, 10.2, 10.3, 10.4, and 10.5 (see Sect. 10.3) are used for designing conical joints.

To facilitate applying the equations, graphs are proposed in this book. In the following subsections and examples, we present the graphs and the process of applying them to solving the relevant problems to the conical joints.

In the design, according to DIN 1448, the value of cone angle, $\alpha = 5.724°$ is usually considered (Orlov 1988; Decker and Kabus 2014).

10.2.1 Method I

In this method, the assumption is that the contact pressure between the shaft and hub does not exceed 100 MPa ($\sigma = 100$ MPa) (In the design, the value of allowable stress is usually considered to be 100 MPa).

Example 10.1 A shaft with diameter of $D = 50$ mm is connected to a gear through a conical joint with a cone angle of $\alpha = 5.724°$ (Fig. 10.2). Find the maximum transmitted torque of the joint.

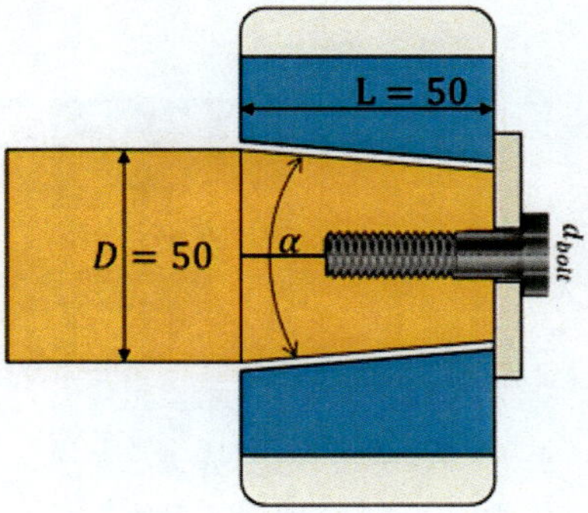

Fig. 10.2 The conical joint of Example 10.1

Solution

On the upper vertical axis of

1. Figure 10.3, we choose $D = 50$ mm, go right to reach the approximate curve of $L = 50$ mm (the connection length is usually considered as $L \leq 1.2D$ (Orlov 1988)), and then go down to achieve the maximum torque that can be transmitted by the joint:

$$T = 1700\ \text{N} \cdot \text{m}$$

2. Now, we look for the bolt size. We keep going down from the last point of the previous step in Fig. 10.3 to cross the approximate cure related to $D = 50$ mm in the fourth quadrant, and then going left to the vertical axis gives us the diameter of bolt:

$$d_{bolt} = 19\text{mm}$$

3. We can round it up to $d_{bolt} = 20$ mm. Therefore, the appropriate bolt for the joint is:

M20 (the property class of 8.8)

Note: Usually in method I, the bolt diameter size is $d_{bolt} \approx 0.4D$.

(In these calculations, the bolt is assumed to be of the property class of 8.8.)

4. To obtain the tightening torque, we find $d_{bolt} = 20$ mm on the lower vertical axis of Fig. 10.3, go left to intersect the curve in the third quadrat, and then go up to the horizontal axis. The tightening torque is:

$$M = 390\ \text{N} \cdot \text{m}$$

5. The smaller diameter of frustum cone is:

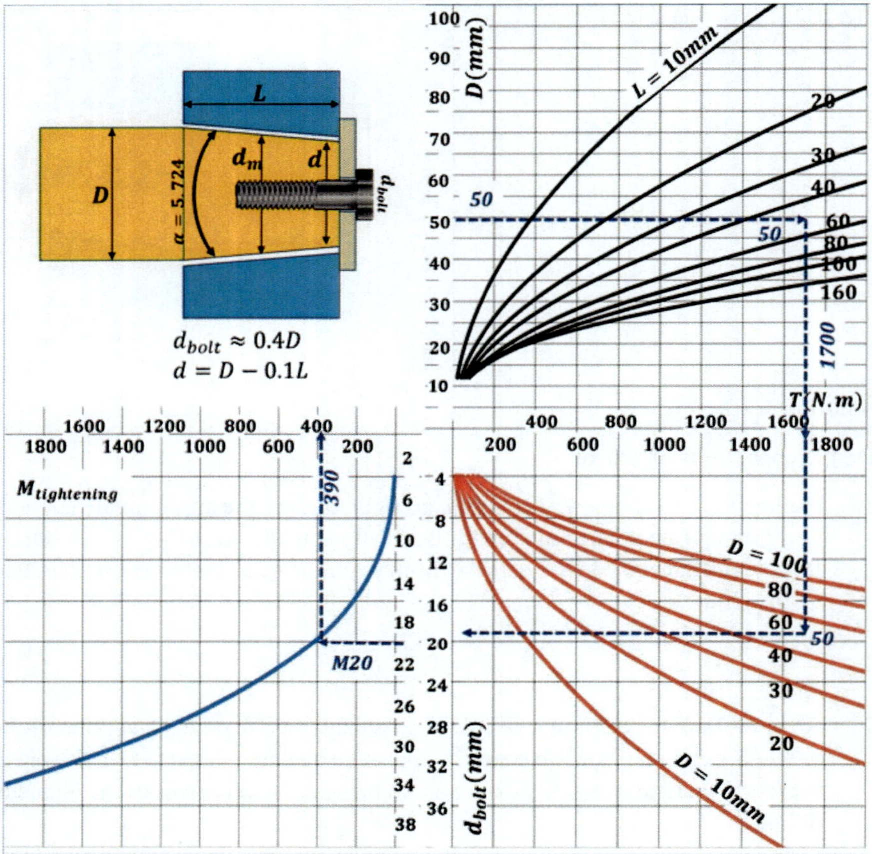

Fig. 10.3 Graphs for designing conical joints (Method I)

$$d = D - 0.1L \xrightarrow[L=50]{D=50} d = 50 - 0.1 \times 50 = 45\text{mm}$$

10.2.2 Method II

The working length (or connection length) is usually assumed to be $D \leq L \leq 1.2D$. In addition, the amount of allowable contact stress (σ) is chosen according to operational conditions of the joint and material of the shaft (and hub) (Table 10.1). This method relies on the mentioned two assumptions.

Example 10.2 Assume that a conical joint ($\alpha = 5.724°$) is supposed to transmit a torque of $T = 800$ N·m from an electromotor. The shaft and hub are made of Ck45 and St37, respectively. Determine the shaft diameter (D).

Table 10.1 The allowable contact stresses for the conical joints (Böge 2011; Orlov 1988; Autodesk Inc. 2021)

	Allowable contact stresses, $\sigma\,(MPa)$				
Shaft material	Autodesk handbook			Orlov	Alfred Böge
	Static loading	Repeated loading	Alternating loading	–	–
Steel grade 37, 42, free-cutting steel, st37, st44	90	63	45	–	$= \frac{\text{yield strength}}{1.5}$
Steel grade 50, St52	125	90	56	–	
Steel grade 60, 28Mn6	160	100	63	100	
Steel grade 70, high-grade and alloy steel	180	110	70	100	
Cast Steel	80	56	40	–	
gray Cast Iron	70	50	32	25	
Bronze	32	22	16	–	

Solution

Between the material of shaft and hub, we select the weaker one (St37). The loading on the shaft can be regarded as on-way (*repeated*) load. However, to be on the safe side, according to Table 10.1, we apply the allowable stress of $\sigma = 45$ MPa (which corresponds to alternating loading).

1. On the horizontal axis of Fig. 10.4, we select T = 800 N·m, go up to cross the curve of $\sigma = 45$ MPa and also the curve of "*M.Y method: Shaft diameter*" (our proposed curve for determining the shaft diameter), and then go left to obtain the corresponding diameters, $D = 50$ and 65 mm, respectively.
2. For transmitting the torque, $T = 800$ N·m, by the conical joint, a diameter of 50 mm for the shaft can be enough. Nevertheless, a diameter of 65 mm is required for the shaft to avoid fatigue failure. Therefore, the final choice is $D = 65$ mm.
3. The bolt diameter is:

$$d_{bolt} \approx 0.2D \xrightarrow{D=65} d_{bolt} \approx 0.2 \times 65 \approx 13\text{mm} \rightarrow \text{M14}$$

4. If a nut is used instead of the bolt (an shown in Fig. 10.4), the size of the nut is:

$$d_{nut} \approx (0.6 - 0.7)D \xrightarrow{D=65} d_{nut} \approx (0.6 - 0.7) \times 65 \approx 39 - 45\text{mm} \rightarrow \text{M42}$$

5. The smaller diameter of frustum cone is:

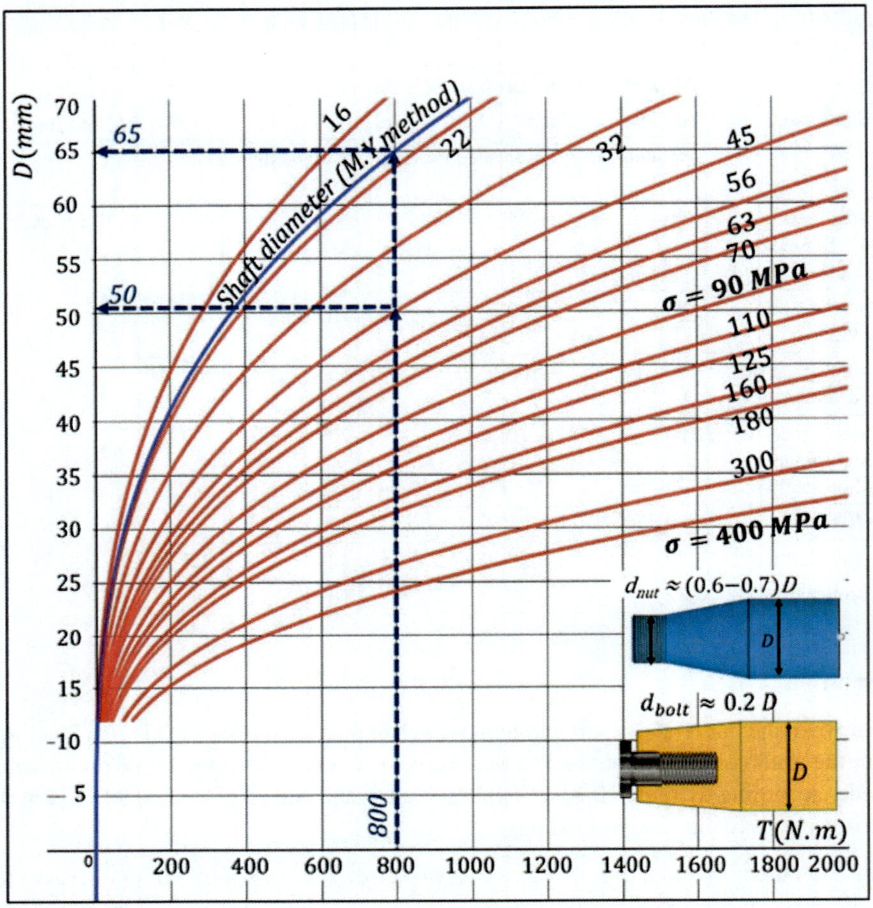

Fig. 10.4 A graph for designing conical joints (Method II)

$$D = 65$$

$$d = D - 0.1L \xrightarrow{if L = 65} d = 65 - 0.1 \times 65 = 58.5 \text{ mm}$$

Note Generally, when there is no vibration or impact, the strengths of conical joints are higher than those of shafts. However, in the operating conditions with vibration (for instance if we connect the shaft of Example 10.2 to a diesel engine instead of an electromotor), the coefficient of friction (which is assumed to be 0.1) may become up to 10 times smaller (i.e. about 0.01) (Klebanov et al. 2007). In which case, the torque should be multiplied by a factor between 8 and 10.

To prevent the reduction of the coefficient of friction, the conical joints are usually used together with square or rectangular keys. This creates a very strong connection.

10.3 Equations

$$T = \frac{\sigma d_m^2 \pi \mu l}{2000 \cos\left(\frac{\alpha}{2}\right)}$$

$$d_m = 0.5(D + d) \approx D\left(1 - \frac{l}{D}\tan\left(\frac{\alpha}{2}\right)\right) \approx 0.95D \tag{10.1}$$

$$F = \frac{2000T}{d_m \mu} \times \frac{\sin\left(\rho + \frac{\alpha}{2}\right)}{\cos(\rho)} \approx \frac{2000T}{d_m \mu} \times 0.154 \tag{10.2}$$

$$F_{pre} \approx 212.27\, d_{bolt}^{2.1149} \tag{10.3}$$

$$M \approx 0.0383\, d_{bolt}^{3.0791} \tag{10.4}$$

$$C = \frac{D - d}{l} = 2\tan\left(\frac{\alpha}{2}\right) \tag{10.5}$$

References

Autodesk Inc. (2021) Engineer's handbook. Accessed from Autodesk Inventor support and learning center: https://knowledge.autodesk.com/

Böge A (2011) Handbuch Maschinenbau: Grundlagen und Anwendungen der Maschinenbau-Technik. Vieweg+Teubner, Wiesbaden (in German)

Decker K-H, Kabus K (2014) Decker Maschinenelemente: Funktion, Gestaltung und Berechnung. Carl Hanser Verlag, München (in German)

Klebanov BM, Barlam DM, Nystrom FE (2007) Machine elements: life and design, 1st edn. CRC Press, Boca Raton

Orlov P (1988) Fundamentals of machine design. Mechanical engineering, Moscow (in Russian)

Chapter 11
Couplings

Nomenclature

D	Coupling size (or outer diameter), mm
d_{shaft}	Shaft diameter, mm
T	Torque, N m

11.1 Introduction

Couplings are components used to connect two shafts at their ends, in order to transmit power and torque between two separate devices. The driving shaft usually transfers rotary motion from a power-generating device such as a motor or engine. While, the driven shaft is commonly the input of a transmission or a driven machine.

In the industry, different couplings are used. Figure 11.1 shows some of them. Flanged couplings and universal couplings (U-joint) are separately discussed in Chaps. 12 and 13, respectively.

The aim of this chapter is to partially compare the transmission of torque by the couplings using a graphical method.

11.2 Design Method

The most important characteristic of the couplings for determining their capacity of torque transmission, is their size (coupling outer diameter, D) (Fig. 11.1). To determine the precise maximum transmitted torque of the couplings, we should refer

M. Yaghoubi and H. Tavakoli, *Mechanical Design of Machine Elements by Graphical Methods*, Materials Forming, Machining and Tribology,
https://doi.org/10.1007/978-3-031-04329-1_11

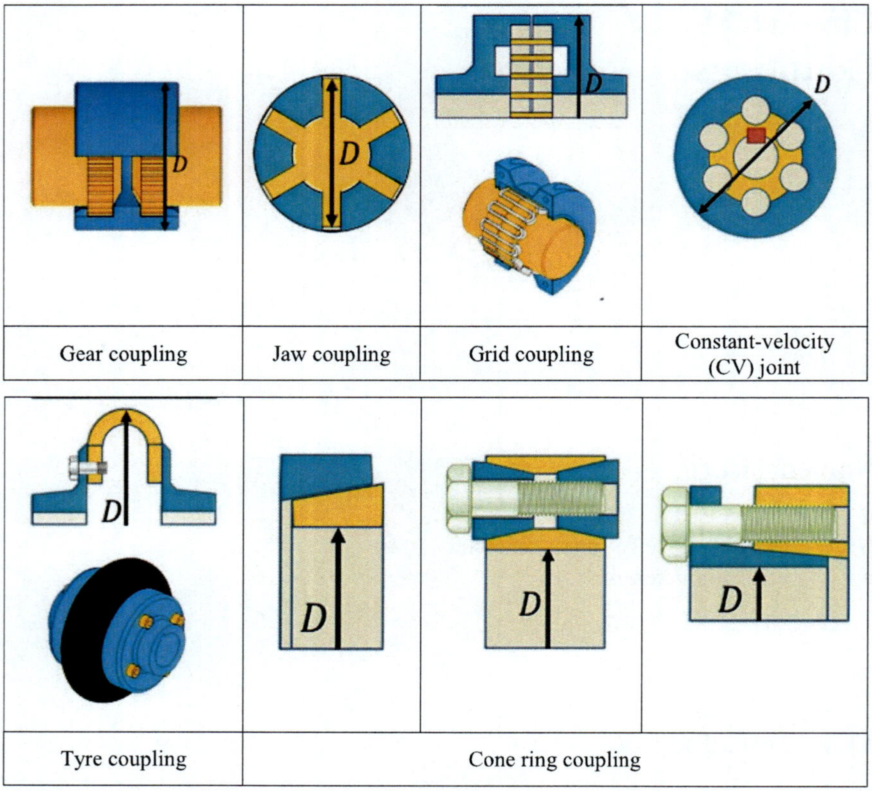

Fig. 11.1 An illustration of some couplings

to manufacturers' catalogs. However, for a rough estimate, we can use Fig. 11.2. The following example illustrates the procedure of applying the method.

Example 11.1 A torque of 1000 N m should be transmitted from shaft of a pump to another shaft. Assume that the type of coupling for connecting the two shafts is Jaw coupling. Determine the shaft diameter (d_{shaft}) and the coupling size (D) (Fig. 11.1).

Solution

1. From the vertical axis of Fig. 11.2, we select the torque, $T = 1000$ N m, and go right to intersect the curve related to *"shaft under pure torsion diagram*" and "*shaft under torsion and bending*", as well as the curve No. 2 (which is for Jaw coupling).
2. Then we go down to find the minimum shaft diameter of the pump (d_{shaft}) and the minimum size of the coupling (D):

   If there is no shaft-misalignment: $d_{shaft} = 50$ mm.

   If there is potential for the shaft-misalignment: $d_{shaft} = 70$ mm.

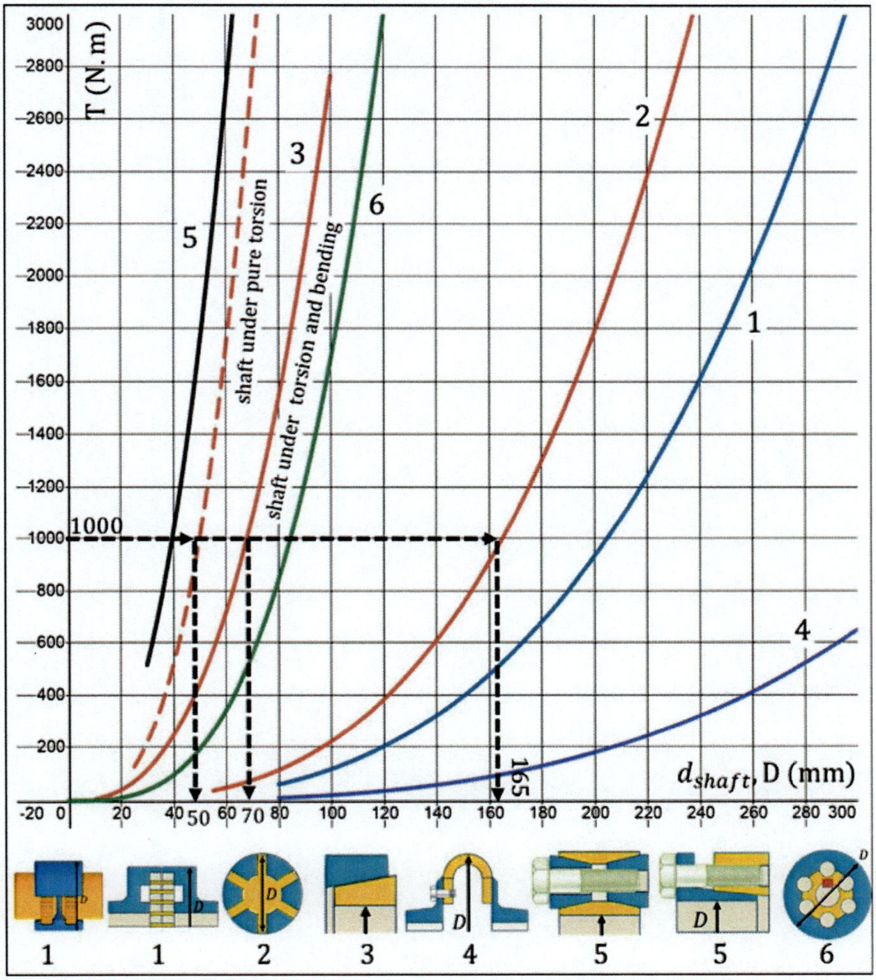

Fig. 11.2 Relationship between transmitted torque and both coupling size and shaft diameter for different couplings

The minimum coupling size (outer diameter): $D = 165$ mm

In conclusion, we should consider a coupling of $D > 165$ mm for this example.

Note: The curve No. 3 is related to a one-pair cone ring coupling. If you use a four-pair cone ring coupling, the transmitted torque will be doubled.

Chapter 12
Flanged Couplings

Nomenclature

A	The effective section area of bolt, mm^2
b	Coupling width, mm
D_2	Diameter of bolt circle, mm
D_3	Hub diameter, mm
d	Shaft diameter, mm
d_2	Bolt diameter, mm
l	Hub length, mm
M	Tightening torque, N m
m	Tightening factor
T	Torque, N m
z	Number of bolts
α	Stress concentration factor
μ	Coefficient of friction ($\mu = 0.1$)
σ_y	Yield strength of bolt, MPa
τ	Allowable shear stress, MPa

12.1 Introduction

Flanged couplings are kind of rigid connections that are used to mate rotating shafts and transmit torque and power between them. They are applied when no relative motion should occur between the driving and driven shafts and an accurate shaft alignment is required. In the coupling, the flanges are connected to the ends of the shafts usually by keys and are attached together using a series of bolts on a bolt circle. The bolts are typically in two types, common (or ordinary) and fit (Fig. 12.1). Indeed,

M. Yaghoubi and H. Tavakoli, *Mechanical Design of Machine Elements by Graphical Methods*, Materials Forming, Machining and Tribology,
https://doi.org/10.1007/978-3-031-04329-1_12

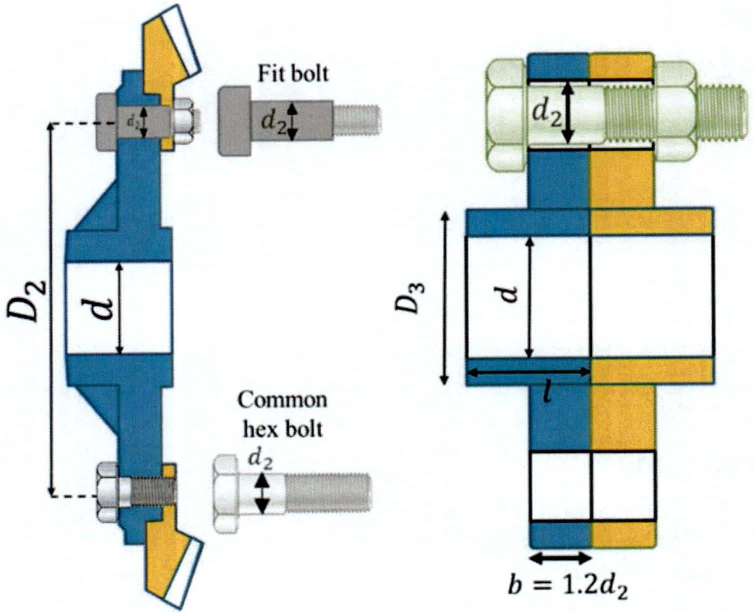

Fig. 12.1 Schematic of a flanged coupling and its notations

the torque or power transmision is performed through the bolts (by the frictional and shear forces).

One of the most important characteristics of flanged couplings is the amount of tightening torque of the bolts. Because the tighter the bolt, the higher the friction, while, the lower the shear strength of the bolts.

In this chapter, a graphical method is proposed for designing flanged couplings in a facilitated manner.

12.2 Design Methods

12.2.1 Method I

The flanged coupling design includes determining the number of bolts (z), bolt diameter (d_2), the diameter of bolt circle (D_2), coupling width (b), hub length (l) and tightening torque of the bolts (M) (see Fig. 12.1). Equations 12.1 and 12.2 can be applied for this purpose (see Sect. 12.3). However, to facilitate the design, we propose graphs of Fig. 12.2, instead of the equations.

We explain the proposed method within the following examples.

Note: It is recommended that the martial of the flanges to be stronger than C35.

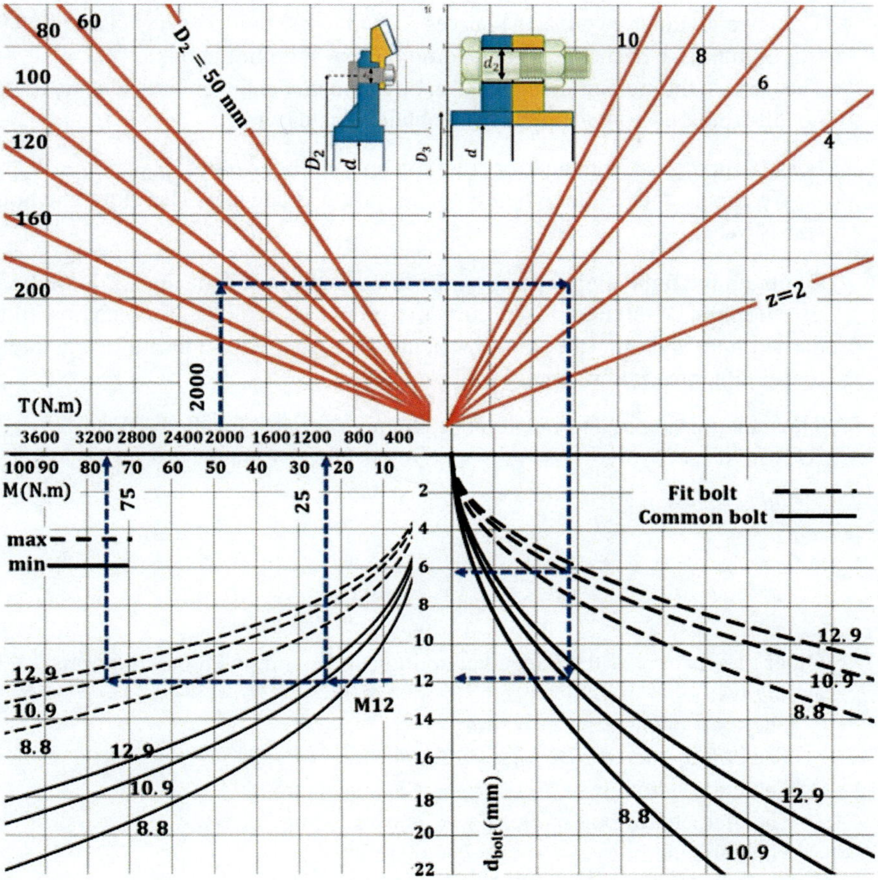

Fig. 12.2 Graphs for the flanged coupling design

Example 12.1 A torque of 1000 N m is supposed to be transmitted between two shafts through a flanged coupling. Determine the coupling's characteristics. The shaft diameter is $d = 60$ mm and the bolt property class is 10.9.

Solution

1. First, for safety, we multiply the torque by a factor of 2:

$$T = 2 \times 1000 = 2000\,\mathrm{N\,m}$$

2. From the left horizontal axis of Fig. 12.2, we choose $T = 2000$ N m, and go up to the curve of $D_2 = 120$ mm (the diameter of bolt circle is usually assumed to be $D_2 \geq 2d$).
3. From the intersection point, we go right to reach the curve of $z = 6$ (we arbitrarily select 6 for the number of bolts).

4. Then, we go down to cross the curves related to the bolt property class of 10.9 (for both fit (the dashed line) and common (the solid line) bolts).
5. The corresponding bolt diameters can be found on the lower vertical axis, if we go left from the crossed points. The obtained values are:

   For fit bolt: $d_2 = 6$ mm and for common bolt: $d_2 = 12$ mm

   We choose the common bolt type, because it is cheaper. Therefore, the bolt size is M12.

6. To obtain the tightening torque, on the lower vertical axis of Fig. 12.2, we find $d_2 = 12$ mm, go left to cross the curves of property class 10.9 (for both max (the dashed line) and min (the solid line)), and then go up to achieve the tightening torques on the left horizontal axis.

   Thus, the minimum and maximum tightening torques for the bolts are 25 and 75 N m, respectively:

$$M_{min} = 25\,\mathrm{N\,m}$$

$$M_{max} = 75\,\mathrm{N\,m}$$

7. Other dimensions of the flanged coupling can be computed based on Fig. 12.3 as follows:

   The shaft diameter: $d = 60\,\mathrm{mm}$.
   The bolt diameter: $d_2(d_{bolt}) = 12\,\mathrm{mm}$.
   The bolt width:b $= (1 - 1.2)d_2 = 12 - 14.4\,\mathrm{mm}$.
   The diameter of bolt circle: $D_2 = 2d = 120\,\mathrm{mm}$.
   The hub length: $l = (1 - 1.3)d = 60 - 78\,\mathrm{mm}$.
   The hub diameter: $D_3 = (1.3 - 1.5)d = 78 - 90\,\mathrm{mm}$.

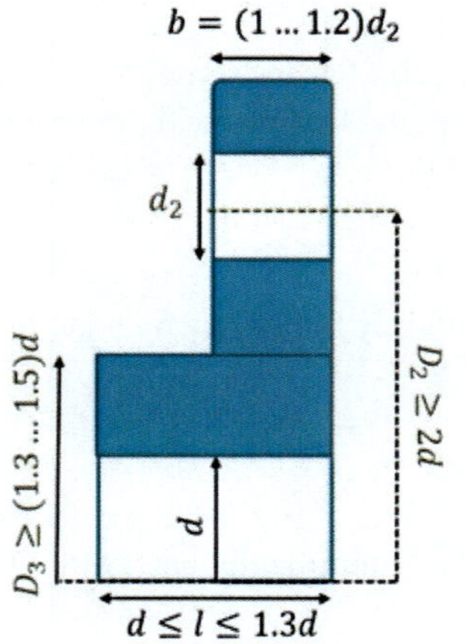

Fig. 12.3 An illustration for recommended dimensions of flanged couplings

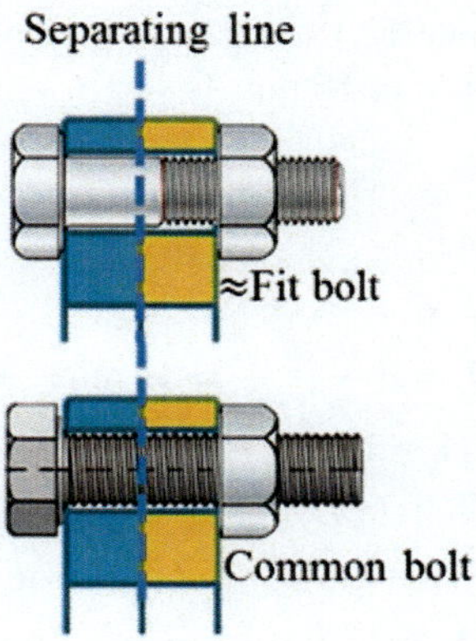

Fig. 12.4 The threaded portion of either of the two bolt types in the separating line of the two flanges

Note: In the case that the common bolt type is chosen for the flanged coupling and the threaded portion is not in the separating line of the two flanges (Fig. 12.4), we can also use fit bolts.

12.2.2 Method II (M.Y Method)

In this method, which is our proposed method in this book, all the dimensions of flanged coupling can be estimated as a function of the shaft diameter (d), as shown in Fig. 12.5. Refer to Example 12.2.

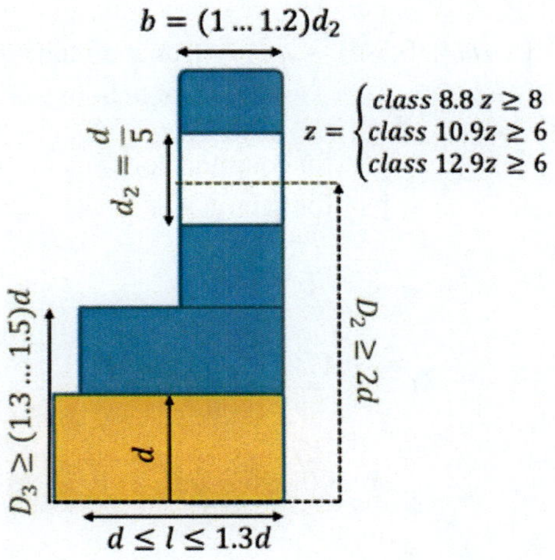

Fig. 12.5 Dimensions of flanged coupling according to M.Y method

Example 12.2 Solve Example 12.1 using M.Y method. The bolt property class is 10.9 and the shaft diameter is $d = 60$ mm.

Solution

1. According to Fig. 12.5 (for common bolt type):
   Number of bolts: $z \geq 6$.
   The bolt diameter: $d_2 = \frac{d}{5} \xrightarrow{d=60} \frac{60}{5} = 12\,\text{mm} \rightarrow$ M12.
   The bolt width: $b = (1 - 1.2)d_2 \xrightarrow{d_2=12} 12 - 14.4\,\text{mm}$.
   The diameter of bolt circle: $D_2 \geq 2d \xrightarrow{d=60} 120\text{mm}$.
   The hub length: $l = (1 - 1.3)d \xrightarrow{d=60} 60 - 78\,\text{mm}$.
   The hub diameter: $D_3 \geq (1.3 - 1.5)d \xrightarrow{d=60} 78 - 90\,\text{mm}$.
2. If we choose the fit bolt type, we should divide the value of d_2 in half:

$$d_2 = \frac{12}{2} = 6\,\text{mm (M6)}$$

3. The tightening torque can be obtained using Fig. 12.2. If you select it from the DIN standard table, it is recommended that you increase the bolt diameter to one higher size (for instance, in this example use M14 instead of M12).

12.3 Equations

$$1000T = 0.5\, D_2\, Az \frac{\tau}{\alpha} \tag{12.1}$$

$$A = \begin{cases} \frac{\pi}{4}(0.87\, d_2 - 0.4288)^2 & \text{(for common bolts)} \\ \frac{\pi}{4} d_2^2 & \text{(for fit bolts)} \end{cases}$$

$$\tau = \begin{cases} (\mu m + 0.5\sqrt[2]{1 - m^2})\sigma_y & \text{(for common bolts)} \\ 0.2\sigma_y & \text{(for fit bolts)} \end{cases}$$

$$\alpha = \begin{cases} 3 \text{ (for common bolts)} \\ 1 \text{ (for fit bolts)} \end{cases}$$

$$\mu = 0.1$$

$$M = \frac{0.17(0.85 m \sigma_y A) d_2}{1000}$$

$$m = 0.196 - 0.6 \tag{12.2}$$

Chapter 13
Universal Joints

Nomenclature

a	Service factor
K	U-joint width, mm
L_h	Theoretical bearing life (or durability), hours
n	Rotational speed of U-joint, rpm
T	Torque, N m
T_{life}	Bearing life factor, N m
β	Operating (or inclination) angle of U-joint, degrees

13.1 Introduction

Universal joints (U-joints), also known as Cardan joints, are flexible couplings used to mate shafts, which are misaligned (at angles greater than 3°). Three types of U-joints are available, namely, single U-joint, double U-joint and constant velocity U-joint (CV joint). Single U-joints have an important drawback that the rotational speed ratio between the output and input shafts is not constant. This may potentially cause torsional vibration problems in the system. If the double U-joint arrangement is used, the speed variations can decrease significantly. Constant velocity ratio U-joints have also been developed that are known as CV joints (Collins et al. 2009; Juvinall and Marshek 2011; Mott et al. 2018).

M. Yaghoubi and H. Tavakoli, *Mechanical Design of Machine Elements by Graphical Methods*, Materials Forming, Machining and Tribology,
https://doi.org/10.1007/978-3-031-04329-1_13

13.2 Design Methods

13.2.1 Relationship Between Torque and U-joint Width

One of the most important characteristics of U-joints is the dimension K (so-called U-joint width) (see Fig. 13.1). The dimension can be used for determining the torque-transmitting capacity of U-joints, and vice versa. Equations 13.1 (Sharipov 2009) and 13.2 (Voith GmbH & Co. KGaA 2020) can be used for this purpose (see Sect. 13.3).

Intead of the equations, the graph shown in Fig. 13.1 can be applied for the estimation. Example 13.1 illustrates the procedure of using the graph.

Example 13.1 Suppose that a U-joint is used for transmitting a torque of 2000 N m from a diesel engine. Determine the size of the U-joint (K).

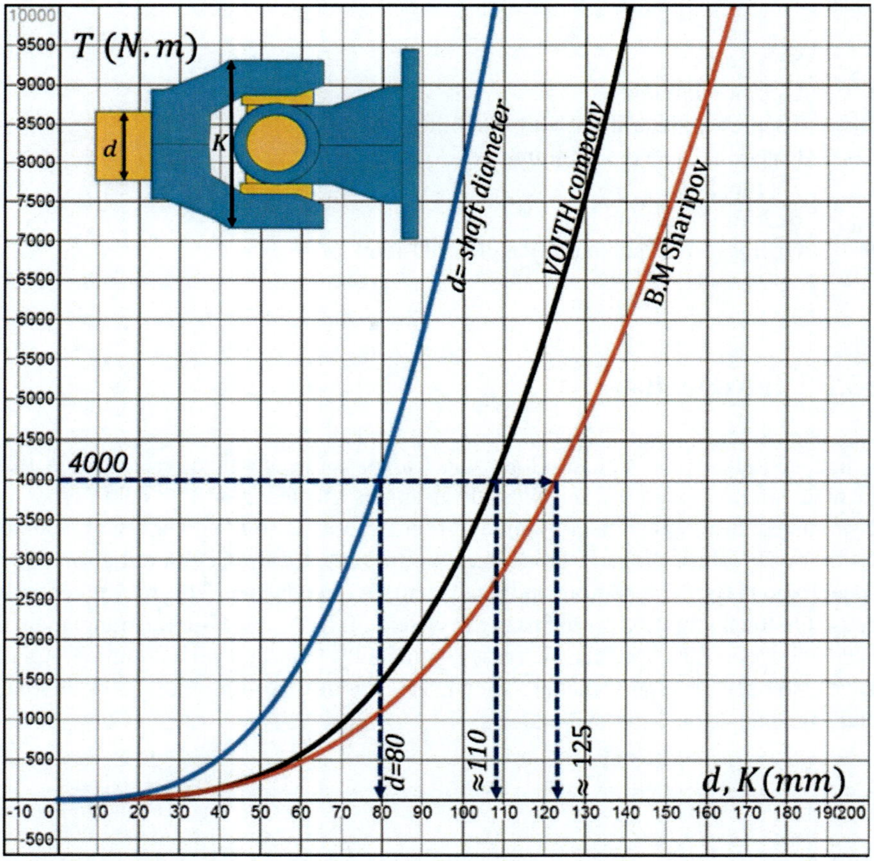

Fig. 13.1 Relationship between the transmitted torque and both the dimension K and shaft diameter of U-joints ($a = 1$ is applied for the curve of *VOITH Company*)

Solution

1. For safety, we multiply the torque by a factor of 2:

$$\mathrm{T} = 2000 \times 2 = 4000 \text{ N m}$$

Note: If you are intended to use the curve of Voith company, select a service factor greater than 1.5 for the pulsating torques (), and a service factor greater than 3 for the alternating torques ().

2. On the vertical axis of Fig. 13.1, we select $T = 4000$ N m, go right to cross both the curves, and then go down to achieve approximate values for K:
   According to the curve related to *Sharipov*: $K \geq 125$mm.
   According to the curve related to *VOITH Company*: $K \approx 110$mm.
3. For more safety, we choose the minimum value of $K = 160$mm.
4. The propeller shaft diameter is also obtained as $d = 80$ mm.

13.2.2 Estimation of the Bearing Life

The life or durability of U-joints is, in fact, the life of their bearings (hereafter, we refer to it as the bearing life). To estimate the life theoretically, we can employ Eqs. 13.3 (Voith GmbH & Co. KGaA 2020) and 13.4 (Kop-Flex Inc 2020) (see Sect. 13.3). To facilitate using the equations, we created graphs of Fig. 13.2. The following example shows how we use the graphs for estimating the bearing life.

Example 13.2 A torque of 2000 N m is supposed to be transmitted at a rotational speed of 1000 rpm by a U-joint (Fig. 13.3). Assume that the inclination angle of the U-joint is $\beta = 3°$ and the U-joint width is $K = 160$ mm (obtained in Example 13.1). Specify the bearing life of the U-joint.

Solution

1. First, from the left horizontal axis of Fig. 13.2, we find $T = 2000$ N m, and go up to intersect the curve of $K = 160$ mm.
2. Then, we go right to reach the curve related to the rotational speed of $n = 1000$ rpm, and go down to cross the curve associated with the inclination angle of $\beta = 3°$.
3. From the crossed point, going right to the lower vertical axis, gives us the bearing life:

$$L_h = 14000 \text{ h}$$

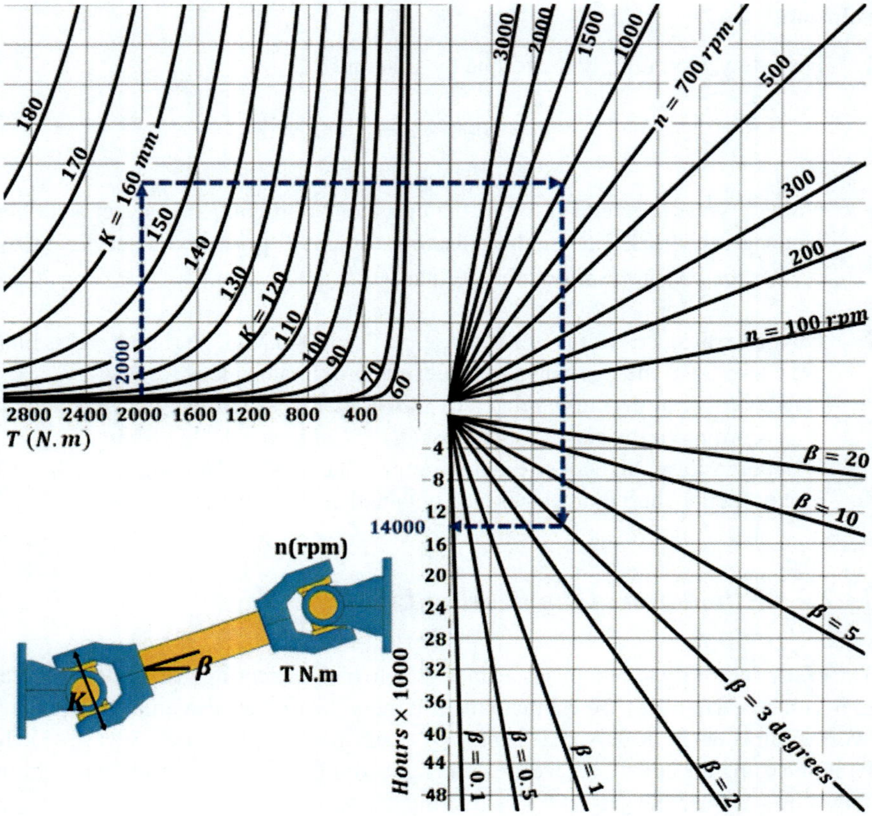

Fig. 13.2 Graphs for estimating the bearing life of U-joints

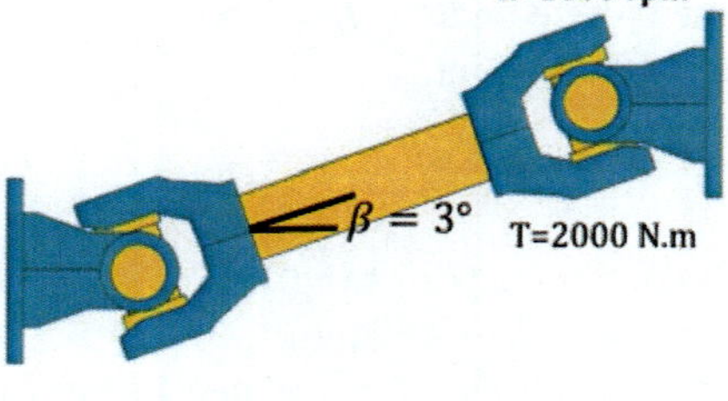

Fig. 13.3 Schematic of the U-joint of Example 13.2

13.3 Equations

$$\mathrm{K} \geq 7.73\sqrt[3]{\mathrm{T}}\ \text{(for tractors)} \tag{13.1}$$

$$K \approx 9.1\ \sqrt[336]{aT}\ \text{(for other equipment)}$$
$$a = \begin{cases} 1 & \text{(for constant loads)} \\ 1.5 & \text{(for pulsating loads)} \\ 3 & \text{(for altemating loads)} \end{cases} \tag{13.2}$$

$$L_h = \frac{1.5 \times 10^7}{n\beta} \times \left(\frac{T_{life}}{T}\right)^{\frac{10}{3}}\ \text{(Voith company)}$$
$$T_{\text{life}} \approx 0.0003K^{3.1586} \tag{13.3}$$

$$L_h = \frac{1.5 \times 10^6}{n\beta} \times \left(\frac{T_{\text{life}}}{T}\right)^{\frac{10}{3}}\ \text{(KopFlex company)}$$
$$T_{\text{life}} \approx 0.00048K^{3.2188} \tag{13.4}$$

References

Collins J, Busby H, Staab G (2009) Mechanical design of machine elements and machines: a failure prevention perspective, 2nd edn. Wiley, New York

Juvinall R, Marshek K (2011) Fundamentals of machine component design, 5th edn. Wiley, New York

Kop-Flex Inc. (2020) Couplings. Retrieved from Regal Rexnord Corporation. Web site: www.regalrexnord.com/Brands/Kop-Flex

Mott RL, Vavrek EM, Wang J (2018) Machine elements in mechanical design, 6th edn. Pearson Education, New York

Sharipov VM (2009) Construction and calculation of tractors: a textbook for university students. Mechanical Engineering, Moscow (in Russian)

Voith GmbH & Co. KGaA (2020) Universal joint shafts. Retrieved from Voith Group Web site: www.voith.com/universal-joint

Chapter 14
Pin Joints

Nomenclature

a	Width of the fork or Clevis in a Clevis pin joint, mm
b	Total thickness at the shear pin location, mm
b_1	Thickness of the sheet at the shear pin location, mm
b_2	Width of the rod at the location of engagement in a Clevis pin joint, mm
b_3	Engagement length of a cantilever pin in the body, mm
D_1	Outer diameter of the reinforcement ring, mm
D_2	Length of the rod at the location of engagement in a Clevis pin joint, mm
D_3	Outer diameter of a hub, mm
d	Shaft diameter, mm
d_p	In diameter, mm
F	Applied force, N
H	A coefficient for mounting style of clevis pin joints
k	A coefficient
L	Clevis length in Clevis pin joints, mm
L_F	Distance between the applied force to cantilever pin and the surface of body, mm
l	The effective (or working) length of pin, mm
m	Ratio of the rod width (b_2) to the pin diameter (d_p) in a Clevis pin joint, ratio of the total fork width (2a) to the pin diameter (d_p) in a Clevis pin joint
m_2	Ratio of the engagement length (b_3) to the pin diameter (d_p) in a cantilever pin
n	Ratio of the total thickness (b) to the pin diameter (d_p) in a shear pin
n_2	Ratio of the distance between the applied force to the pin and the surface of body (L_F), to the pin diameter (d_p) in a cantilever pin
p	Allowable bearing stress, MPa
p_{hub}	Bearing stress of hubs, MPa
p_{shaft}	Bearing stress of shafts, MPa

M. Yaghoubi and H. Tavakoli, *Mechanical Design of Machine Elements by Graphical Methods*, Materials Forming, Machining and Tribology,
https://doi.org/10.1007/978-3-031-04329-1_14

Q	A coefficient for estimating the tolerable force in cantilever pin
R_m	Tensile strength of pins, MPa
T	Torque, N·m
t	Thickness of the reinforcement ring, mm
t_2	Difference between radii of the hub and shaft in a radial pin joint
t_3	Thickness of a hub, mm
t_p	Thickness of a hollow pin, mm
σ_b	Allowable bending stress, MPa
τ	Allowable shear stress, MPa
τ_{pin}	Shear stress of pins, MPa

14.1 Introduction

Pins are simple and inexpensive components used to join and fasten machine elements together. They are usually employed when low to medium torques are to be transmitted (Jiang 2019). Different types of pins with different functions are available. However, in this chapter we only discuss shear pins, Clevis pins and Dowel pins.

A shear pin acts as a safety or protective component.

A Clevis joint consists of two pieces, a U-shaped piece (known as Clevis) and a Clevis pin. The Clevis has holes at the end of the prongs for inserting the Clevis pin. The Clevis pin usually has a domed head at one end and a cross-hole at the other end. A cotter pin is inserted into the hole at the end of the pin to keep the Clevis pin in place.

A Dowel pin is simply a cylinder of solid material. The pin joins the mating elements either radially (radial pins), or axially (axial pins) or in a manner like a cantilever beam (cantilever pins).

Pins are generally subjected to bending, crushing, and shearing stresses. All these stresses must be taken into account for their design. This chapter deals with designing pin joints using our proposed graphs.

14.2 Design Methods

Allowable stresses for different material of pins under different loads are presented in Table 14.1. The values are used for designing pins.

14.2.1 Shear Pin

Figure 14.1 shows an arrangement of a shear pin that is used for joining sheets. This

Table 14.1 Allowable stresses for pin joints under different loads

DIN (DIN743)	Decker (Decker and Kabus 2014)	Allowable stresses* *(under alternating loads)*	
$0.1R_m$	≈ 30	τ (MPa)	
$0.15R_m$	70	σ_b (MPa)	
20	12	Movable fit	p (MPa)
$\approx 0.15R_m$	≈ 36	Immovable fit	

Autodesk (Autodesk Inc. 2021)

p (MPa)						Part material	τ (MPa)			σ_b (MPa)			Pin material
Movable fit			Immovable fit				Movable fit			Immovable fit			
A	R	C	A	R	C		A	R	C	A	R	C	
12	65	30	50	24	84	Steel grade 37, 42	25	35	50	35	55	80	11 373, 11 423, 11 110
24	90	30	90	24	120	Steel grade 50, 60—high grade and alloy steels	35	80	70	50	50	110	11 500
12	60	30	40	24	80	Cast steel	42	95	85	60	60	130	11 600
16	50	40	30	32	70	Gray cast iron	48	110	100	68	68	150	11 700, 12 040

* τ, σ_b, p, R_m, C, R and A stand for allowable shear stress, allowable bending stress, allowable bearing stress, tensile strength of pin, Constant load, Repeating load, and Alternating load, respectively

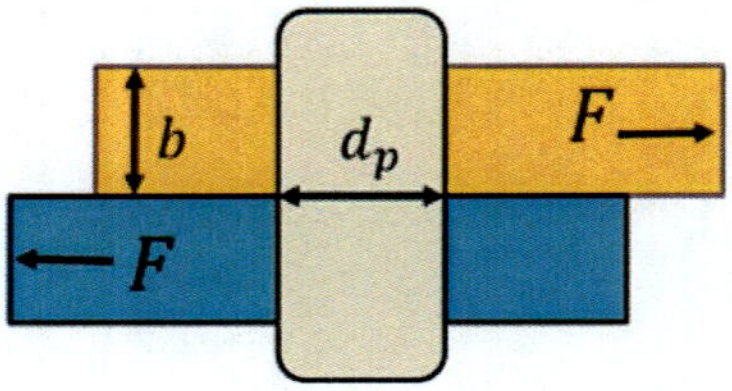

Fig. 14.1 Schematic of a shear pin joint

type of pins is under shearing and crushing stresses. Equation 14.1 can be applied for design of shear pins (see Sect. 14.3). Instead, graphs in Figs. 14.2 and 14.3 can be used as a simpler method. For more details about applying the graphs for problems related to shear pins' design, refer to Example 14.1.

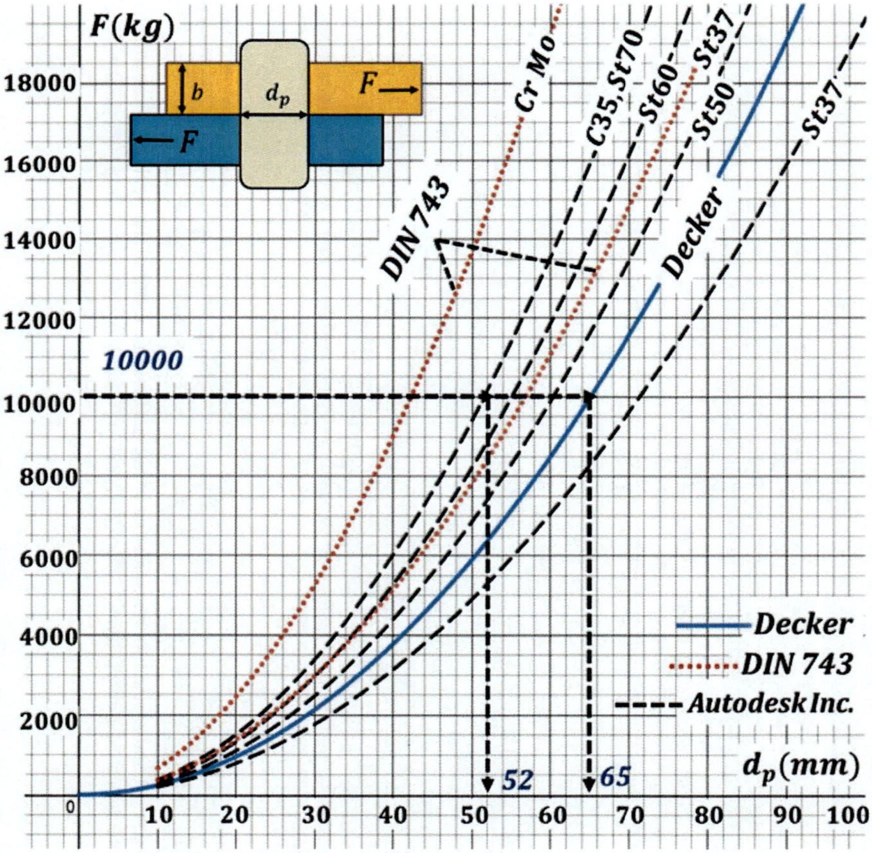

Fig. 14.2 A graph for determining the shear pin diameter

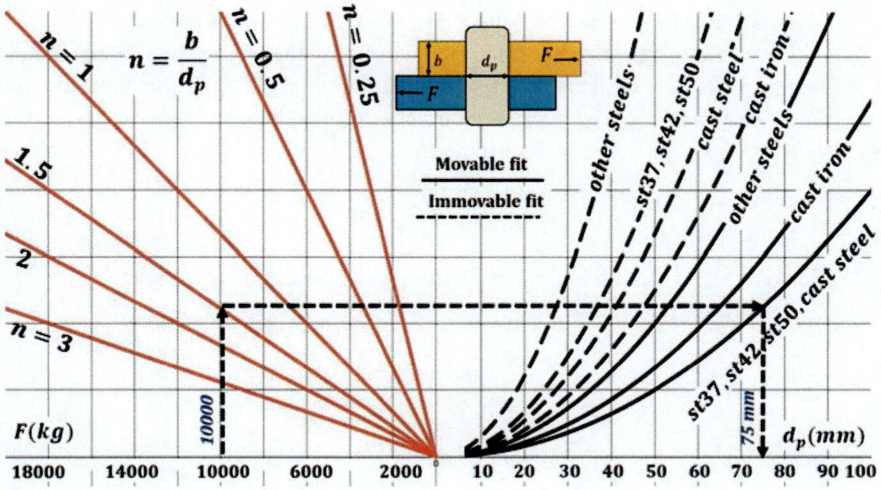

Fig. 14.3 Graphs for determining the hole diameter in shear pin joint

Example 14.1 In a hydraulic press machine, shear pins like the one shown in Fig. 14.1, are used to connect side sheets to the frame. The sheets are made of St37 with a thickness of $b = 60$ mm. Assume that material of the pins is C35 and each pin is under a force of 10 tons (10,000 kg). Estimate the pin size (diameter) and the required thickness of the sheets at the pin location (b).

Solution First, we check the strength of the pin using Fig. 14.2. Graphs based on three methods, namely *Decker* (Decker and Kabus 2014), *DIN 734*, and *Autodesk Inc.* (Autodesk Inc. 2021), are presented in this figure.

1. From the vertical axis, we select the force $F = 10{,}000$ kg and go right to intersect the curve of C35. Then, we go down to determine the pin size:

$$d_p = 52\,\text{mm}$$

   If it does not matter what the pin material is, or if it is unknown, we can use the curve of *Decker*. In this case, the obtained value for the pin size is:

$$d_p = 65\,\text{mm}$$

   Now we should check the strength of the sheet in the place of the hole.

   For easy assembly, we fit the pin into the hole slowly (not pressed). Thus, the pin in the hole can slide (movable fit).

2. In Fig. 14.3, we find $F = 10{,}000$ kg on the left horizontal axis and go up to cross the curve of $n = 1.5$.

   We suppose that the thickness of sheet is as large as 1.5 times of the pin diameter ($b = 1.5d_p$).

3. Then, we go right to intersect the curve of *st37* (the solid line).
   Between the pin and sheet material, we choose the weaker one. Her the sheet material (st37) is weaker than the pin material (C35). In addition, since the pin fit is "*movable fit*", we choose the curve with solid line.
4. After that, we go down to determine the pin diameter:

$$d_p = 75\,\text{mm}$$

5. Between $d_p = 52$ mm and $d_p = 75$ mm, we choose the bigger one.
   Therefore, the results are:

$$d_p = 75\,\text{mm}$$

$$b = nd_p = 1.5 \times 75 = 112.5\,\text{mm}$$

The thickness of the sheet is 60 mm, however, the required thickness is $b =$ 112.5 mm. Therefore, we must weld a *reinforcement ring* with outer diameter of D_1 and thickness of t around the hole of the sheet as shown in Fig. 14.4a, to increase the sheet thickness at that place to 112.5 mm.

The dimensions of the reinforcement ring are:

$$\text{D} = (2 - 2.5)d_p \xrightarrow{d_p = 75\,\text{mm}} \text{D} = 150 - 187.5\,\text{mm}$$

$$t = b - b_1 \xrightarrow[b_1 = 60\,\text{mm}]{b = 112.5\,\text{mm}} \text{t} = 112.5 - 60 = 52.5\,\text{mm}$$

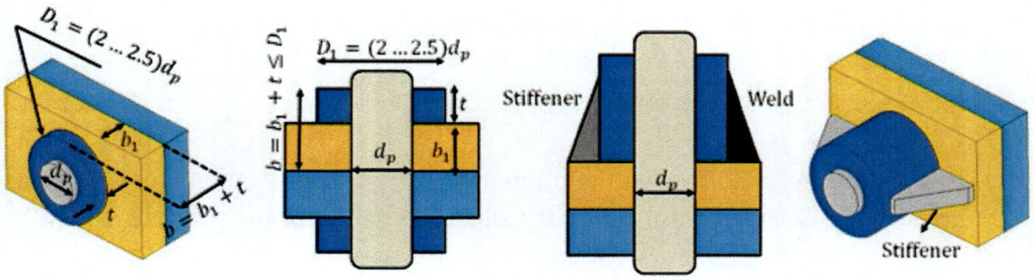

Fig. 14.4 Increasing the thickness of sheet using reinforcement ring, **a** reinforcement ring together with the pin and sheet **b** weld and triangular stiffener for increasing the stiffness of the ring

Notes:

- The sheet thickness should be less than or equal to the outer diameter of the reinforcement ring ($b \leq D_1$). If $b > D_1$, either increase D_1 or increase the weld thickness around the ring or place triangular stiffeners around it to increase the stiffness of the ring, as shown in Fig. 14.4b.
- If you use a grooved pin, multiply the diameter of the shear pin by 1.2.
- Sometimes designers use a bolt instead of a pin. In this case, if you tighten the bolt manually, use the curve CrMo in Fig. 14.2; however, if you tighten it according to the DIN standard table, use the Decker curve or the curve St37 of Autodesk Inc. (because preload weakens the bolt).

14.2.2 Clevis Joint

Figure 14.5 shows schematic of a Clevis joint. In the following subsections, two methods are introduced for designing this kind of joint.

14.2.2.1 General Method

In this method, Eqs. 14.2 and 14.3 (see Sect. 14.3) can be used for designing Clevis joints. However, we propose graphs of Figs. 14.6 and 14.7 for easier designing. More details can be found in the following example.

Example 14.2 A clevis joint together with specification of its geometry is schematically shown in Fig. 14.8. Find the maximum load (F) that the joint can tolerate. Assume that material of the pin, rod (the orange piece) and U-shaped piece (Clevis or fork) are Ck45, Ck45 and st52, respectively.

Solution For finding the maximum permissible force, we have to check the pin in terms of bending and crushing and the rod and Clevis in terms of crushing.

Fig. 14.5 Schematic of a Clevis pin joint

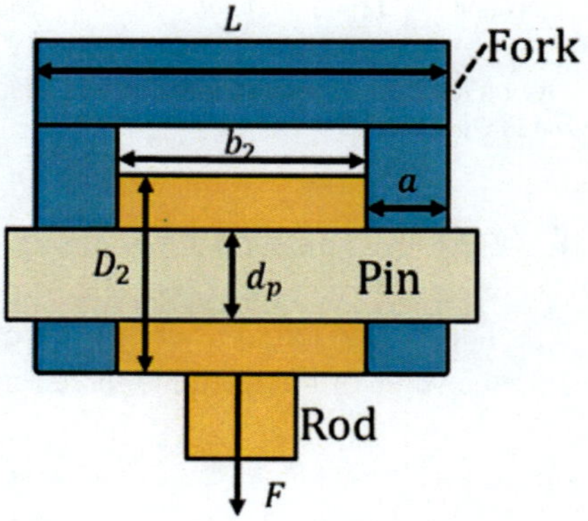

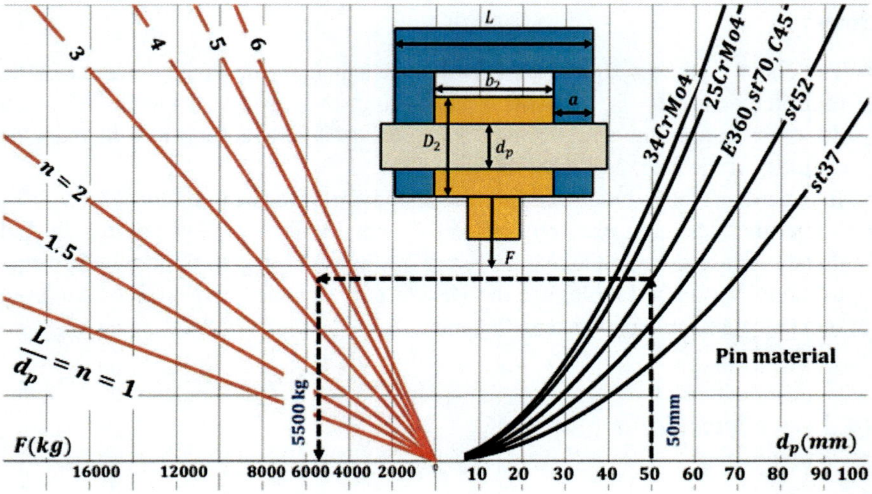

Fig. 14.6 Graphs for estimating the maximum tolerable force of Clevis pins under bending

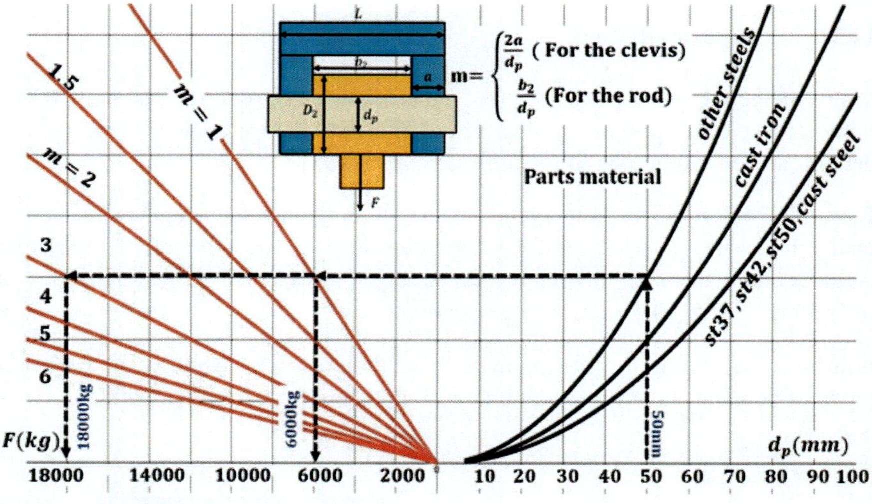

Fig. 14.7 Graphs for estimating the maximum tolerable force of the U-shaped piece (Clevis) and rod in Clevis joints

1. First, from the right horizontal axis of Fig. 14.6, we locate $d_p = 50$ mm and go up to intersect the curve related to the pin material *Ck45*.
2. Then we go left to reach the curve of $n = \frac{L}{d_p} = \frac{200}{50} = 4$, and go down to obtain the value of the tolerable force for the pin in bending:

$$F = 5500\,\text{kg}$$

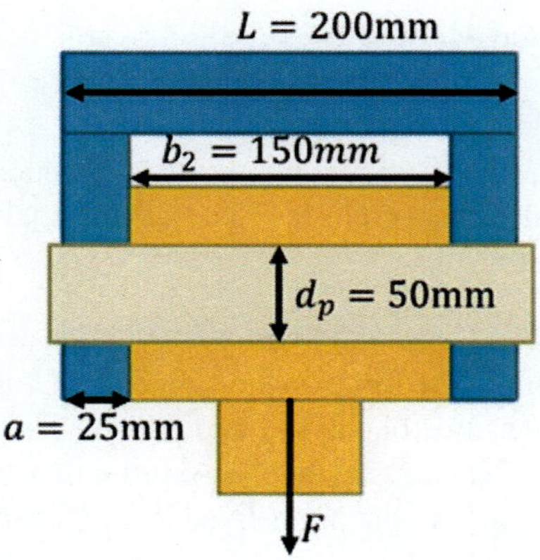

Fig. 14.8 The Clevis joint of Example 14.2

Now, we should calculate the tolerable force for the rod and the Clevis considering their crushing strength.

3. From the right horizontal axis of Fig. 14.7, we select the pin diameter $d_p = 50$ mm and go right to intersect the curve of "*other steels*". Because material of the rod and Clevis (Ck45 and st52, respectively) are in the category of "*other steels*" (Fig. 14.7).
4. Then, we go left to cross the curves related to $m = \frac{b_2}{d_p} = \frac{150}{50} = 3$ for the rod and $m = \frac{2a}{d_p} = \frac{2\times 25}{50} = 1$ for the Clevis.
5. From both the curves, we go down to obtain the maximum tolerable force of the rod and Clevis:

$$F = 18000\,\text{kg}\ \ (\text{for the rod})$$

$$F = 6000\,\text{kg}(\text{for the Clevis})$$

From the three obtained forces, we select the smallest one. Thus, the tolerable force of the Clevis joint is:

$$F = 5500\,\text{kg}$$

Notes:

- In Fig. 14.7, if the pin mounting is "immovable fit", multiply the obtained force (F) by 2.
- Sometimes designers use bush of Cu or Al alloys between the part and pin for prevention of wearing. In this case, from Fig. 14.7 consider st37 or cast iron as the parts material.

14.2.2.2 Recommended Method

In this method, the design is based on Eq. 14.4 (see Sect. 14.3) and the relationships presented in Table 14.2, which are recommended in practical manufacturing.

To facilitate the calculations, graphs in Figs. 14.9 and 14.10 can be applied, instead of the equation and relationships.

Example 14.3 Figure 14.11 shows a hydraulic cylinder, which is connected to the lower plate by a clevis joint. The force on the cylinder is 6 tons (6000 kg). Determine size of the pin and dimensions of the rod and Clevis. The pin, rod and Clevis are made of E360, E360 and St52, respectively. Assume that the pin mounting type is "loose" (not pressed fitting) and the pin is fixed axially (non-sliding).

Solution First, we should check the bending strength of the pin.

1. In Fig. 14.9, from the left horizontal axis, we select the force $F = 6000$ kg and go up to intersect the curve related to the pin material *E360*.

Table 14.2 Recommended relationships between dimensions of Clevis joints, and the coefficient H

Relationship		Parameters
Non-sliding	Sliding*	
d_p	$1.6d_p$	b_2
$0.5d_p$	$0.6d_p$	a
$(2.5-3)d_p$		D_2

* Sliding: the pin is not fixed axially; Non-sliding: the pin is fixed axially

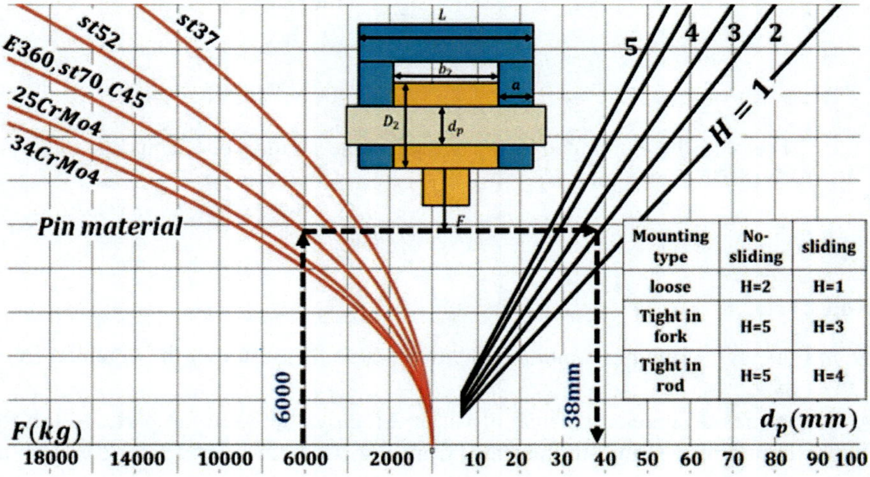

Fig. 14.9 Graphs for estimating maximum bending force of Clevis pins

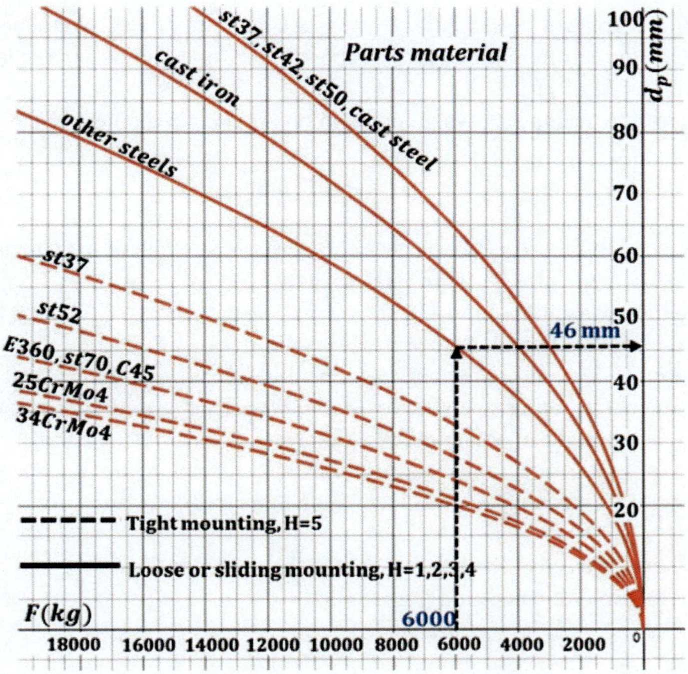

Fig. 14.10 A graph for estimating maximum crushing force of the Clevis and rod in Clevis joints

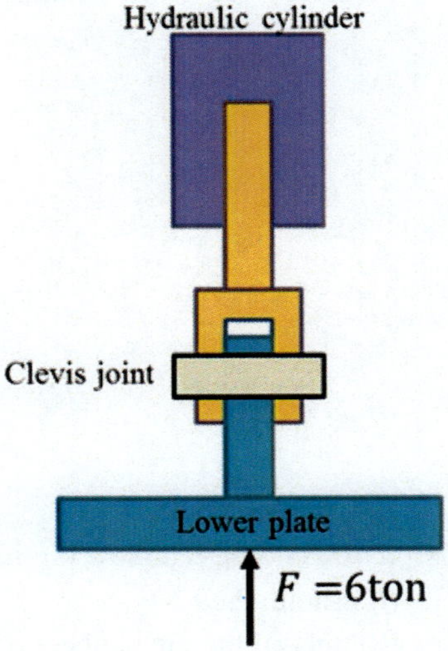

Fig. 14.11 An illustration for Example 14.3

2. Then, we go right horizontally to cross the curve of $H = 2$. We choose $H = 2$ because the pin mounting type is "loose" and "non-sliding" (see the table on the right side of Fig. 14.9).
3. From the crossed point, we go down to obtain the pin diameter:

$$d_p = 38\,\text{mm}$$

Now using Fig. 14.10, we check the crushing strength of the Clevis and rod, which are made of st52 and E360, respectively.

4. From the horizontal axis of Fig. 14.10, we select the force $F = 6000$ kg and go up to reach the curve of "other steels" (solid line).

Note: In Fig. 14.10, the curves with solid line are for "loose" and "sliding" fits. The curve "other steels" is for steels, which are stronger than st50.

5. Then, we go right to obtain the pin diameter:

$$d_p = 46\,\text{mm}$$

6. In the end, we select the largest size for the pin from Figs. 14.9 and 14.10, i.e. $d_p = 46$ mm. We can round it up to:

$$d_p = 50\,\text{mm}$$

7. We can calculate the other dimensions regarding the Clevis joint according to Table 14.2 for "non-sliding" fit as follows:

$$d_p = 50\,\text{mm}$$

$$b_2 = d_p = 50\,\text{mm}$$

$$a = 0.5\,d_p = 0.5 \times 50 = 25\,\text{mm}$$

$$D_2 = (2.5 - 3)d_p = 125\text{--}150\,\text{mm}$$

Notes:

- If you choose a hollow pin instead of a solid pin, the thickness of the hollow pin (t_p) should be $t_p \geq \frac{d_p}{6}$.
- Rigidity of the pin connections is very important, especially for big elements and machines. By increasing rigidity, bending stress of pin joints decreases and consequently you can select pins with smaller diameter (Orlov 1988). For example, as shown in Table 14.3, the bending stress in the arrangement type f is four times less than that in the type a.

Table 14.3 Bending stress ratios for different joint arrangement designs

Joint arrangement		Stress ratio	Joint arrangement		Stress ratio
a		1	*d*		0.5
b		0.75	*e*		0.5
c		0.56	*f*		0.25

- Sometimes designers use bush of Cu or Al alloys between the part and pin for prevention of wearing. In this case, from Fig. 14.10 consider st37 or cast iron (loose or sliding mounting) as the parts material.

Therefore, the calculations above are only for the joint arrangement type *a*. If you design the rod and Clevis with an arrangement like the type *f*, the bending strength of the pin will increase by 4 times. However, the crushing strengths of the rod and Clevis remain unchanged.

14.2.3 Dowel Pin

14.2.3.1 Radial Pin

Radial pins (Fig. 14.12) are used when low torques are to be transmitted. They are designed based on strength of pin, shaft, and hub (shear stress in pin and crushing stress in shaft and hub). Equation 14.5 can be applied for this purpose (see Sect. 14.3) (Fig. 14.12).

For designing radial pins, you can easily employ the graph in Fig. 14.13. In Fig. 14.13 allowable bearing stress (p) and shear stress (τ) are considered according to Decker and Kabus (2014) (Table 14.1).

Example 14.4 A radial pin is used to join a gear and a shaft with diameter of d = 70 mm. The pin's diameter is considered to be $d_p = 0.25\,d \approx 18\,\text{mm}$. The mounting type is *immovable fit* (for both the shaft and the hub). Determine the torque transmitting capacity of the joint.

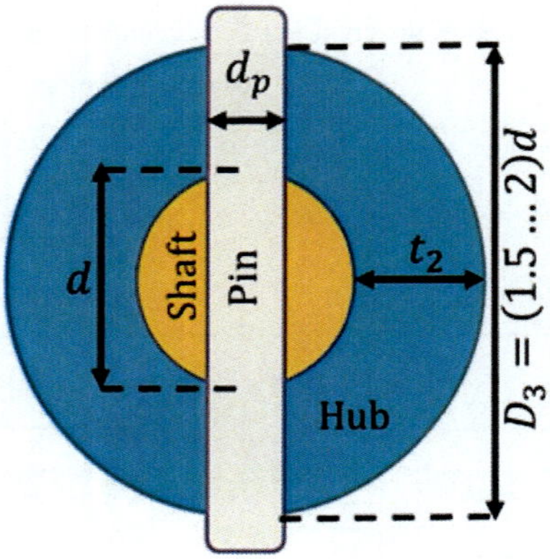

Fig. 14.12 Schematic of a radial pin joint

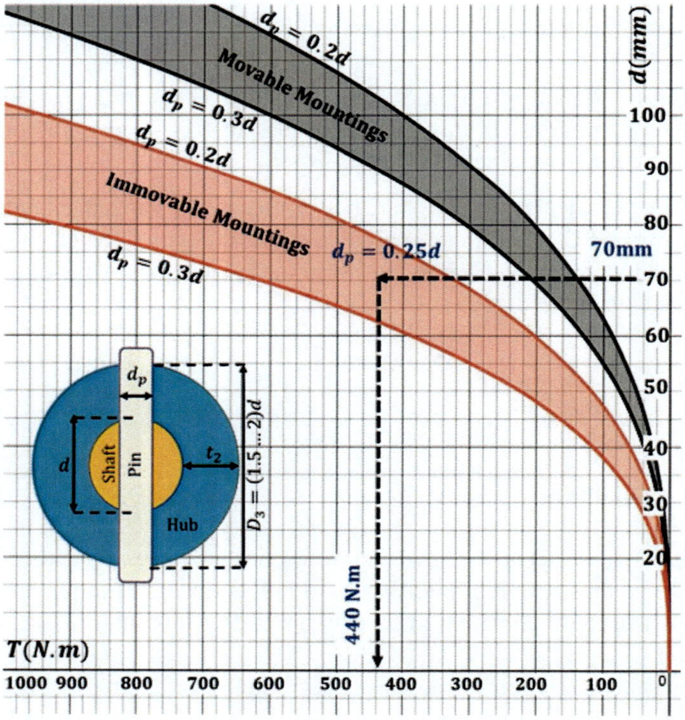

Fig. 14.13 A graph for design of radial pins

Solution

1. First, in Fig. 14.13, we select the shaft diameter $d = 70$ mm from the vertical axis and go left.
2. In the "*immovable*" area, we select the approximate pin diameter $d_p = 0.25\,d \approx 18\,\text{mm}$ and then we go down to determine the maximum transmitted torque:

$$T = 440\,\text{N}\cdot\text{m}$$

   Therefore, the allowable torque to be transmitted by this connection is $T = 440$ N·m.
3. The outer diameter of the hub (D_3), according to Eq. 14.5 (see Sect. 14.3) is:

$$D_3 = (1.5 - 2)d\text{mm}D_3 = 105 - 140\text{mm}$$

   The material of the shaft, pin and hub must be stronger than St37.

Notes:

- If you choose a groove pin instead of a simple Dowel pin, multiply the torque by 0.7. Therefore, for Example 14.4, in case of using a groove pin, the maximum torque would be $440 \times 0.7 = 308$ N·m.
- In connections that radial pins are used for levers such as wrenches, door handles, etc. in which the crushing of the hub or shaft is not important, you can consider the mounting as "*immovable*".

14.2.3.2 Axial Pin

In this type of connection, pins are installed axially between shaft and hub (Fig. 14.14). The design is based on the crushing stress in the shaft and hub. Equation 14.6 (see Sect. 14.3) can be applied for this purpose.

For simplicity of calculations, we present the graph of Fig. 14.15. Since in the literature, different relationships between the pin diameter and the shaft diameter

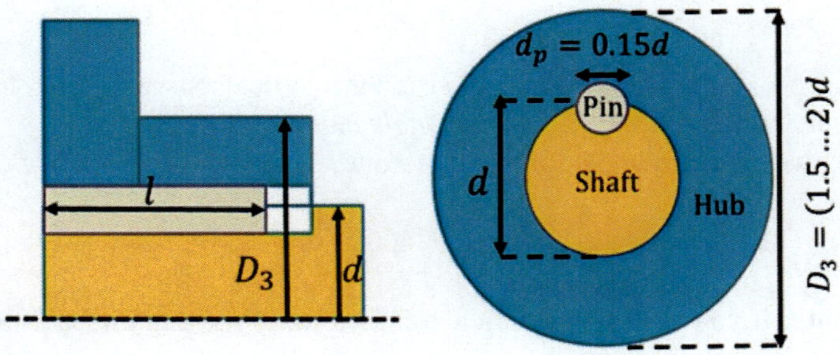

Fig. 14.14 Schematic of an axial pin joint

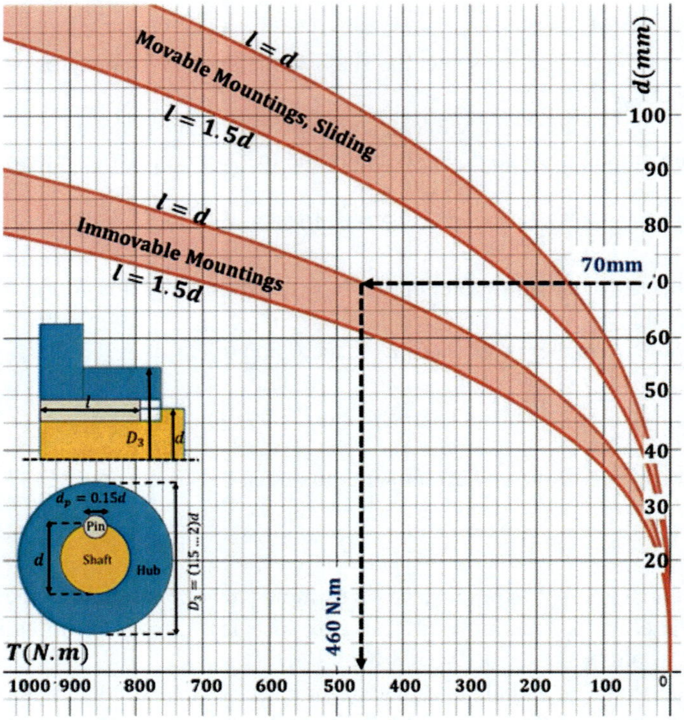

Fig. 14.15 A graph for design of axial pins

have been considered, we take an average one, i.e. $d_p = 0.15\,d$. The allowable bearing stress (p) and shear stress (τ) are derived from Decker and Kabus (2014) (Table 14.1).

Example 14.5 A shaft with diameter of $d = 70$ mm is jointed to a gear using an axial pin. The pin length is $l = d = 70$ mm. The fitting style is *immovable fit* (for both the shaft and the hub). Determine the maximum torque that can be transmitted by the joint.

Solution

1. In Fig. 14.15, we select $d = 70$ mm from the vertical axis and go left to intersect the curve related to $l = d$ in "*immovable mountings*" area.
2. Then, we go down to determine the torque:

$$T = 460\,\mathrm{N \cdot m}$$

Note: If you need to transmit a greater torque, you can increase the number of pins. For instance, if you use two numbers of pins, the transmitted torque will be double.

3. Other dimensions of the pin joint are:

$$d_p = 0.15\,d = 0.15 \times 70 = 105\,\mathrm{mm}$$

$$l = d = 70\,\mathrm{mm}$$

$$D_3 = (1.5 - 2)d = 105 - 140\,\mathrm{mm}$$

Notes:

- For the axial joint, the material of pin, shaft and hub must be stronger than St37.
- If you select a groove pin instead of a simple Dowel pin, multiply the torque by 0.7. Therefore, for Example 14.5, if a groove pin is used, the maximum torque will be 460 × 0.7 = 322 N·m.

14.2.3.3 Cantilever Pin

In design of cantilever pins (Fig. 14.16), the aim is usually to calculate the tolerable force of the pin. The pin is mainly under bending stress and the body is more under crushing stress. The calculations regarding the design can be made using Eq. 14.7 (see Sect. 14.3). Through the following example, we show that instead of the equation, the graph of Fig. 14.17 can be applied.

Example 14.6 In the pin clutch shown in Fig. 14.18, the pin diameter is $d_p = 20$ mm. The pins are mounted in part A by press fitting and in part B as sliding fit. Find the tolerable force of each pin.

1. First from Fig. 14.17, we find the maximum force that each pin can withstand under bending. According to our experience about such connections, L_F can be calculated as:

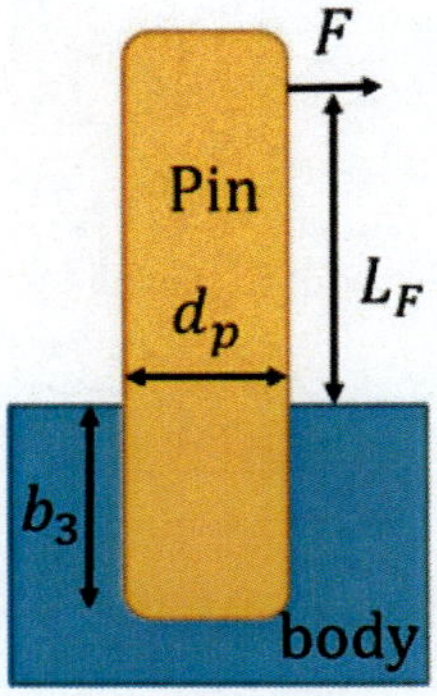

Fig. 14.16 Schematic of a cantilever pin joint

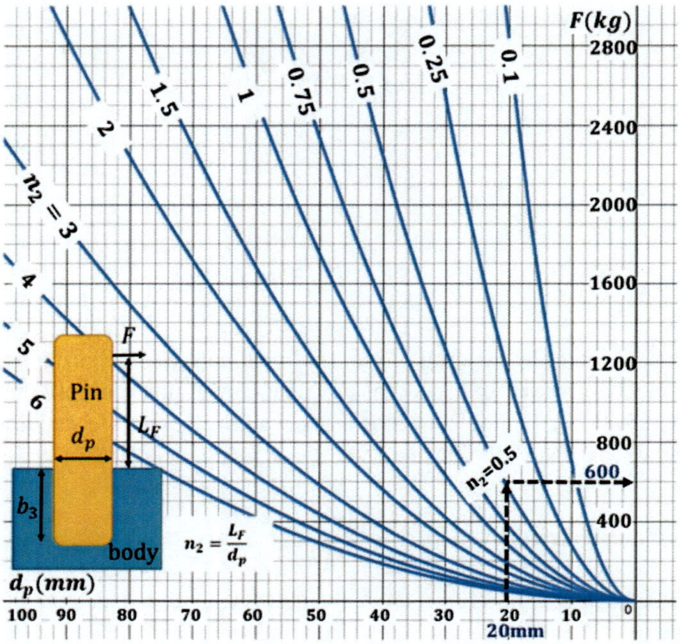

Fig. 14.17 A graph representing the tolerable force (under bending) as a function of the pin diameter for cantilever pins

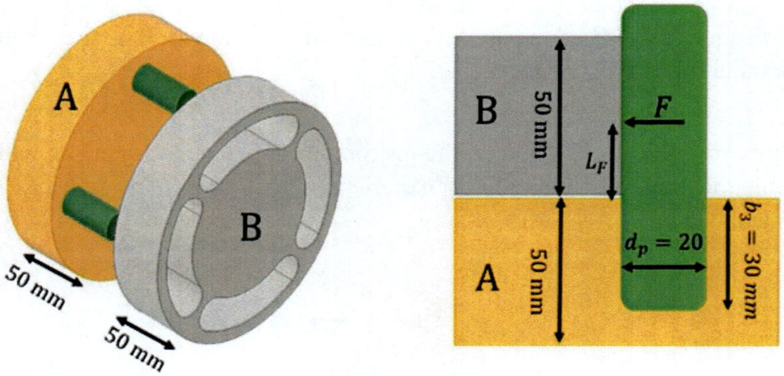

Fig. 14.18 An illustration of a pin clutch for Example 14.6

$$L_F = \frac{width\,of\;part\,\mathrm{B}}{5} = \frac{50}{5} = 10\,\mathrm{mm}$$

2. From the horizontal axis of Fig. 14.17, we select the pin diameter $d_p = 20$ mm and go up to intersect the curve related to $n_2 = \frac{L_F}{d_p} = \frac{10}{20} = 0.5$.
3. Then, we go right to obtain the tolerable force of the pin:

$$F_{pin} = 600\,\text{kg}$$

Now we should determine the maximum forces under which the parts A and B are not crushed. To do this, we need to multiply the force obtained for the pin, $F_{pin} = 600$ kg, by a coefficient (Q) that is obtained from Eq. 14.8 (see Sect. 14.3).

4. Applying Eq. 14.8 yields the following values for the tolerable forces of the parts A and B.

$$n_2 = \frac{L_F}{d_p} = \frac{10}{20} = 0.5$$

$$m_2 = \frac{b_3}{d_p} = \frac{30}{20} = 1.5$$

For part A in which the pins are mounted by press-fitting:

$$Q = \frac{n_2 \times m_2^2}{6n_2 + 4m_2} \times 5.1 \xrightarrow[m_2 = 1.5]{n_2 = 0.5} Q = 0.63$$

$$F_A = F_{pin} \times Q = 600 \times 0.63 = 378\,\text{kg}$$

For part B in which the pins are mounted as sliding fit:

$$Q = \frac{n_2 \times m_2^2}{6n_2 + 4m_2} \times 1.7 \xrightarrow[m_2 = 1.5]{n_2 = 0.5} Q = 0.21$$

$$F_B = F_{pin} \times Q = 600 \times 0.21 = 126\,\text{kg}$$

5. In the end, considering the three obtained forces, $F_{pin} = 600$ kg, $F_A = 378$ kg and $F_B = 126$ kg, the tolerable force of **each pin** should be (we select the minimum value):

$$F = 126\,\text{kg}$$

Note: If more than one pin is used in the clutch, then the force should be multiplied by the number of pins.

14.3 Equations

$$F = bd_p p \xrightarrow{b=nd_p} F = nd_p^2 p$$

$$F = bd_p p \rightarrow^{b=nd_p} F = nd_p^2 p \quad (14.1)$$

(For simplicity, we assume the thickness of sheet to be $b_1 = nd_p$)

$$d_p^3 = \frac{FL}{0.8 \times 0.15\, R_m} \quad \text{(For bending)} \quad (14.2)$$

$$L = b_2 + 2a$$

$$d_p^2 = \frac{F}{mp} \quad \text{(for bearing)}$$

$$\mathrm{m} = \begin{cases} \frac{b_2}{d_p} & \text{(For the rod in Clevis joints)} \\ \frac{2a}{d_p} & \text{(For the Clevis in Clevis joints)} \end{cases} \quad (14.3)$$

$$d_p \approx k\sqrt{\frac{F}{0.15 R_m}}$$

$$d_p = \frac{F}{2ap} \quad (14.4)$$

k coefficient	Assembly	Sliding	Non-sliding
	Loose	1.9	1.6
	Tight in fork	1.4	1.1
	Tight in rod	1.2	1.1

$$p_{hub} = \frac{1000\,T}{d_p t_2 (d + t_2)}$$

$$p_{shaft} = \frac{6000\,T}{d_p d^2} \quad (14.5)$$

$$\tau_{pin} = \frac{4000T}{\pi d_p^2 d}$$

$$d_p = (0.2 - 0.3)d$$

$$t_2 = (0.25 - 0.5)d$$

$$D_3 = (1.5 - 2)d$$

$$p_{shaft} \text{ or } p_{hub} = \frac{4000\,T}{d\,d_p l}$$

$$l = (1 - 1.5)d \tag{14.6}$$

$$d_p = (0.15 - 0.2)d \text{ or } (0.13 - 0.16)d$$

$$D_3 = (1.5 - 2)d$$

$$F = \frac{0.1\,d_p^3}{L_F}\sigma_b$$

$$F = \frac{d_p b_3^2}{6L_F + 4b_3}p \tag{14.7}$$

(σ_b and p according to Table 14.1)

$$Q = \frac{n_2 \times m_2^2}{6n_2 + 4m_2}(1.7 \text{ or } 5.1)$$

(For sliding mounting: 1.7, and for tight or press-fit mounting: 5.1)

$$n_2 = \frac{L_F}{d_p} \tag{14.8}$$

$$m_2 = \frac{b_3}{d_p}$$

References

Autodesk Inc. (2021) Engineer's handbook. Retrieved from Autodesk Inventor support and learning center: https://knowledge.autodesk.com/

Decker K-H, Kabus K (2014) Decker Maschinenelemente: Funktion, Gestaltung und Berechnung. Carl Hanser Verlag, München (in German)

Jiang W (2019) Analysis and design of machine elements, 1st edn. Wiley, New Jersey

Orlov P (1988) Fundamentals of machine design. Mechanical engineering, Moscow (in Russian)

Chapter 15
Shafts and Their Associated Elements All Together

Nomenclature

D	Outer diameter of a hub, mm
D_2	Diameter of the bolt circle, mm
d	Shaft diameter, mm
d_2	Bolt diameter, mm
d_p	Pin diameter, mm
p	Allowable bearing stress, MPa
l	The engagement or working length in joints, mm
M	Bending moment, MPa
T	Torque, N.m
Z	The width of weld, mm
z	Number of bolts
Δ	Interference, mm
σ	Allowable bending stress, MPa
τ	Allowable shear stress, MPa

15.1 Introduction

The common method for designing a machine, is that each component is designed separately, and then they are assembled. Especially for the power transmitting elements, which are associated with each other, this is a very time consuming process.

In this chapter, we propose a method to design and calculate shafts and their attached components all together. The method is on the basis that if the shaft diameter is known, some of the associated elements such as welding, splines, keys, joints, couplings etc. can be designed accordingly.

M. Yaghoubi and H. Tavakoli, *Mechanical Design of Machine Elements by Graphical Methods*, Materials Forming, Machining and Tribology,
https://doi.org/10.1007/978-3-031-04329-1_15

15.2 Design Method (M.Y Method)

In this method, first we determine the shaft diameter of the weakest connection (*base shaft diameter*), then we design the rest of the components according to the shaft size and using Table 15.1. The elements in Table 15.1 has been ordered based on their strength (descending). Therefore, "press-fit joint" (No. 1) is the strongest, and

Table 15.1 The proposed relationships for designing shafts and their attached elements by M.Y method (all dimensions are in mm)

No. 1	No. 2	No. 3	No. 4	No. 5
Press-fit joint 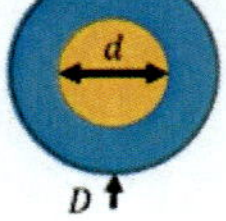	Two side weld (under pure torsion) 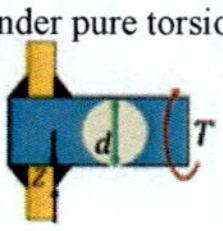	Involute spline  Conical joint 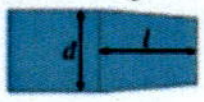	Straight spline (heavy duty) 	Cone ring coupling type A 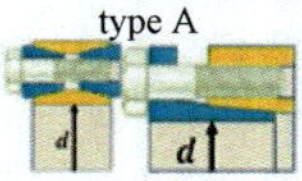
$l \geq d, D \geq 2d$ $\Delta \approx 0.001428d$	$Z \geq 0.2d$	$l \geq d$	$l \geq d$	----
No. 6	No. 7	No. 8	No. 9	No. 10
Flange coupling* $z = 10$ 	Shaft (pure torsion), Chain coupling 	One side weld (under pure torsion) 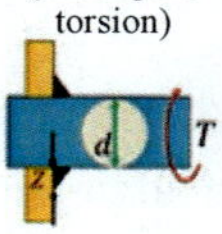	Flange coupling $z = 8$ 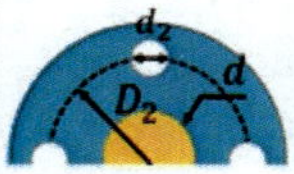	Straight spline (medium duty) 
$d_2 = \frac{d}{5}, D_2 \geq 2d$	$d \geq 5\sqrt[3]{T}$	$Z \geq 0.2d$	$d_2 = \frac{d}{5}$ $D_2 \geq 2d$	$l \geq d$
No. 11	No. 12	No. 13		No. 14
Shaft (bending & torsion) 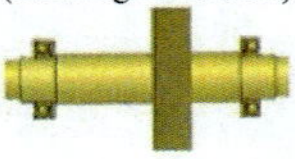	Flange coupling $z = 6$ 	Two side weld (under bending & torsion) 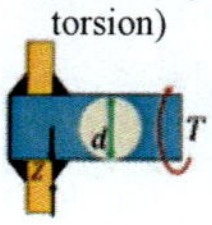	Straight spline (light duty) 	Rectangular key 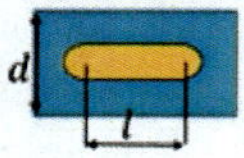
$d \geq 5.5\sqrt[3]{T}$	$d_2 = \frac{d}{5}, D_2 \geq 2d$	$Z \geq 0.2d$	$l \geq d$	$l \geq d$
No. 15		No. 16	No. 17	
Cone ring coupling type B 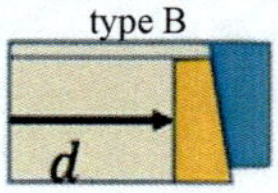	Axial pin 	Shaft (bending & torsion) 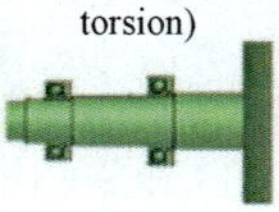	Radial pin 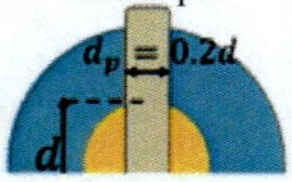	One side weld (under bending & torsion) 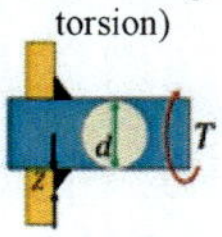
---	$d_p = 0.2d, l \geq d$	$d \geq 7\sqrt[3]{T}$	$d_p = 0.15d$	$Z \geq 0.2d$

*For flange couplings, the bolt property class is 8.8.

"radial pin" and "one-way welding under bending and torsion" (No. 17) are the weakest elements (Table 15.1).

For determining the *base shaft diameter*, Fig. 15.1 can be used.

For instance, for a press-fit connection, if its shaft diameter is known, there is now need to design it using the relevant formula. Instead, the relationships presented in Table 15.1 can be simply applied:

The engagement or connection length (mm): $l \geq d$
Hub diameter (mm): $D \geq 2d$
Interference (mm): $\Delta \approx 0.001428d$

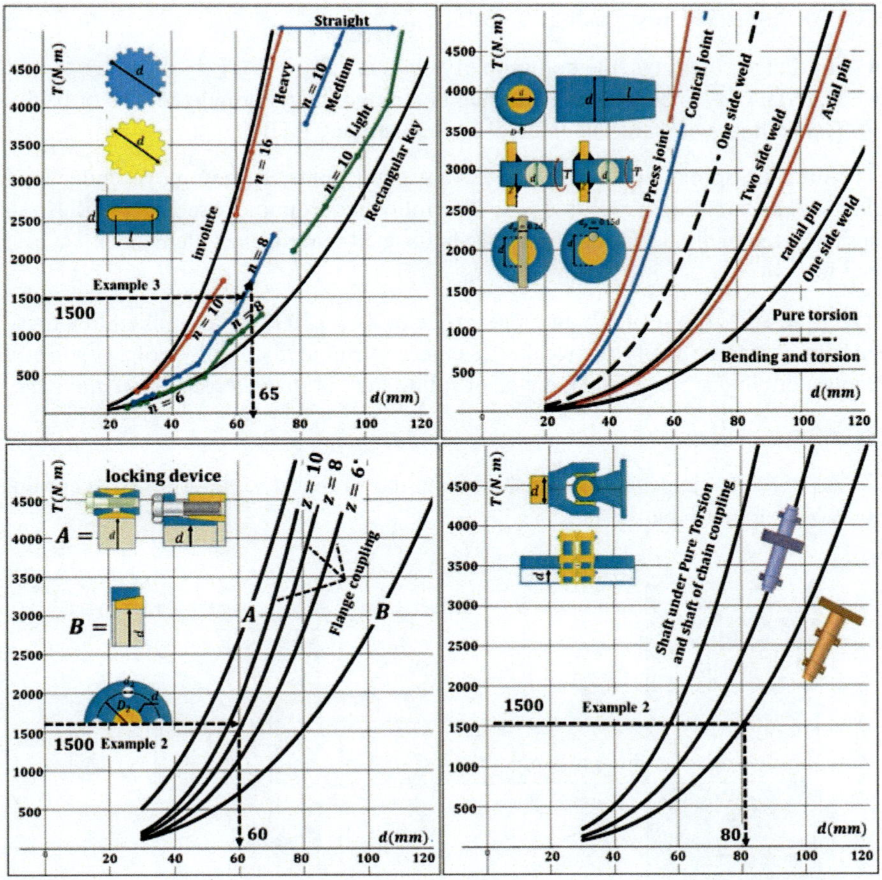

Fig. 15.1 Graphs for specifying the base shaft diameter and determining dimensions of the attached elements to shafts

As another instance, if a pure tension shaft with diameter of d (No. 7 in Table 15.1) is connected to a flange coupling, there is no need to design the flange. Rather, just a flange stronger than the part No. 7 should be chosen, i.e. the flange coupling No. 6:

Number of bolts: $z = 10$
Bolt diameter: $d_2 = \frac{d}{5}$
Diameter of the bolt circle: $D_2 \geq 2d$

Note that if a weaker flange coupling (for example the flange No. 9 in Table 15.1) is selected, the transmission torque of it will be less than that of the shaft. Therefore, it will fail sooner under high torques.

Considerations regarding using Table 15.1:

- For flange couplings, the bolt property class is 8.8.
- Material of elements should be equal or stronger than normalized C35 or C45.
- The hardness of spline shaft should be more than 50 HRC.

Note: the spline shafts in Fig. 15.1 are fixed joints. If they move under load, they will be as strength as radial pins. The following examples are given to illustrate applying the method for designing the shafts and their attached elements in an easy and fast way.

Example 15.1 In Fig. 15.2, an input torque of $T = 1500$ N.m is applied to a transmission system. Specify the weakest element. Assume that number of bolts for the flange coupling is $z = 8$, and the type of spline shaft is *straight teeth – medium duty*.

Solution

We arrange the elements of each shaft from the strongest to the weakest according to M.Y method (Table 15.1).

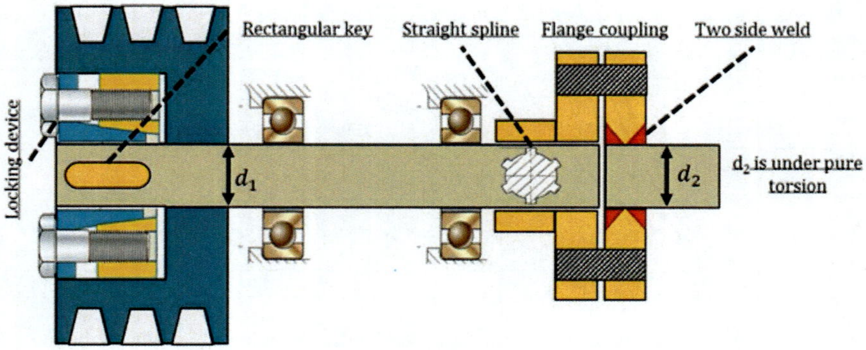

Fig. 15.2 An illustration for Example 15.1

Table 15.2 The separated shafts of Example 15.1 and their associated components

Shaft No. 1 and its attached components:	Shaft No. 2 and its attached components:
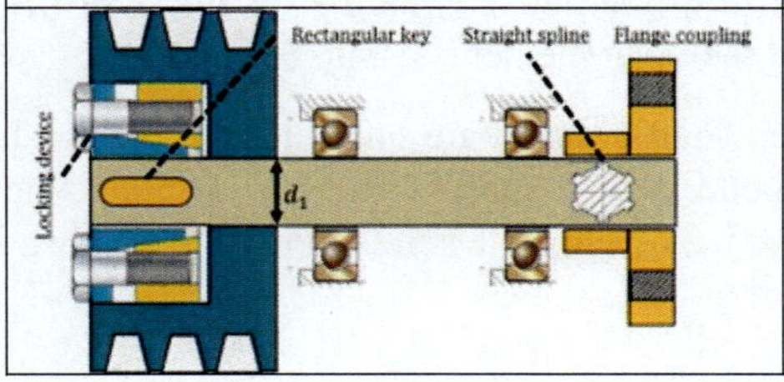	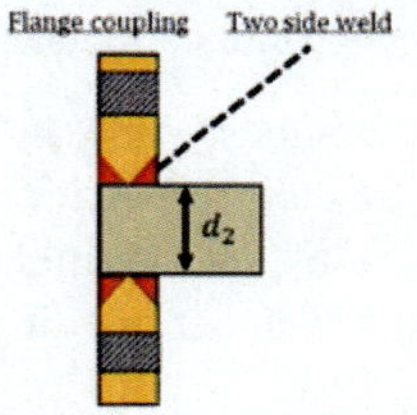

Table 15.3 arrangement of the elements for the shaft No. 1 of Example 15.1

Elemants' order according to Table 15.1				
The strongest	→			The weakest
No. 5	No. 9	No. 10	No. 14	No. 16
Cone ring coupling type A 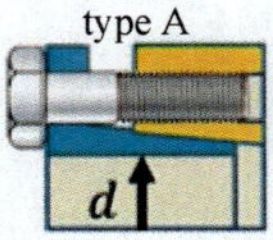	Flange coupling with $z = 8$ 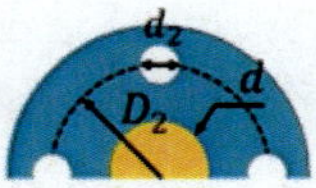	Straight spline (medium duty) 	Rectangular key 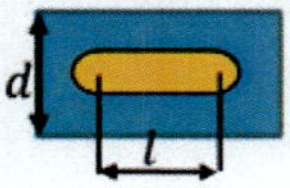	Shaft (bending & torsion) 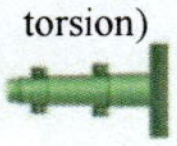

1. First we separate each shaft and its attached components as shown in Table 15.2:
2. Then, we arrange the components from the strongest to the weakest according to Table 15.1.
3. Since the shaft No. 1 is cantilever and is attached to a v-belt, it is under torsion and bending (No. 16 in Table 15.1). Therefore, arrangement of the elements of the shaft No. 1 from the strongest to the weakest is as presented in Table 15.3.
4. The shaft No. 2 is under *pure torsion*. Thus, the arrangement of the elements of the shaft from the strongest to the weakest is as shown in Table 15.4.

Table 15.4 Arrangement of the elements for the shaft No. 2 of Example 15.1

Elemants' order according to Table 15.1		
The strongest	→	The weakest
2	7	9
Two side weld (under pure torsion) 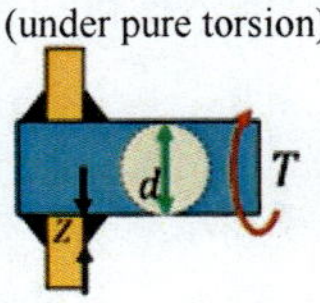	Shaft (pure torsion) 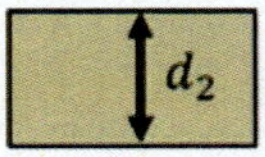	Flange coupling with $z = 8$ 

In conclusion, for the transmission system, the weakest component on the shaft No. 1 is the shaft itself and on the shaft No. 2 is the flange coupling. Therefore, if we have the size of d_I and the shaft size of the flange coupling (as *the base shaft diameters*), we can estimate the size of the other elements using Table 15.1.

Note: In machines, you can rarely find shafts, which are under pure torsion. The examples of their application are propeller shafts, shafts connected to pumps and centrifugal fans, etc. (Wittel et al. 2013). Therefore, it is recommended to always consider shafts under bending and torsion.

Example 15.2 Determine all dimensions of the elements in Example 15.1.

If the values of the base shaft diameters (d_I) and the shaft diameter of the flange coupling) in Example 15.1 were known, we would directly calculate the dimensions of the other elements according to Table 15.1. However, the sizes are not given. In this case, we can find the sizes using Fig. 15.1.

1. According to Example 15.1, the weakest element on the shaft No. 1 is the shaft itself, which is under bending and torsion (). Thus, from the vertical axis of the graph in the fourth quadrant (the lower right corner) of Fig. 15.1, we choose $T = 1500$ N.m and go right to intersect the curve related to the shaft .
2. Then, we go down to find the value of the base shaft diameter for the shaft No. 1:

$$d = 80\,\text{mm}$$

3. The weakest element on the shaft No. 2 is the flange coupling with $z = 8$. Thus, from the vertical axis of the graph in the third quadrant (the lower left corner) of Fig. 15.1, we select $T = 1500$ N.m and go right to cross the curve of $z = 8$.
4. Then, we go down to obtain the value of the base shaft diameter for the shaft No. 2:

$$d = 60\,\text{mm}$$

5. Now by knowing the base shaft diameters, we can calculate the dimensions of the elements on shaft No. 1, as presented in Table 15.5.
6. Moreover, the dimensions of the elements on shaft No. 2 are shown in Table 15.6.
7. As the two flanges must be coupled to each other, we should choose one of the sizes. In order to avoid assembly problems, we chooze the larger one, i.e. the flange of shaft No. 1.

Table 15.5 dimensions of the elements attached to the shaft No. 1 of Example 15.2

The dimensions according to the base shaft diameter No. 1 (d = 80 mm) and Table 15.1

Cone ring coupling type A 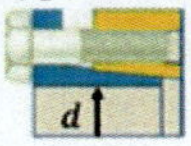	Flange coupling with $z = 8$ 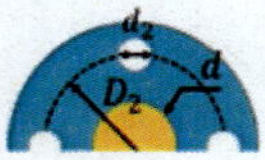	Straight spline (medium duty) 	Rectangular key 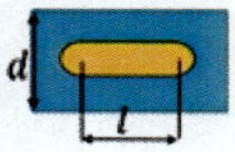	Shaft (bending & torsion) 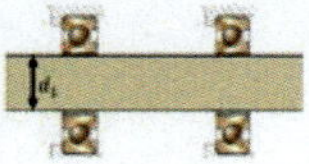
No need to calculate it. By referring to the catalog, select a cone ring coupling type A with a diameter of d = 80 mm	The shaft diameter: $d = 80$ mm the bolt size: $d_2 = \frac{d}{5}$ $\xrightarrow{d=80} d_2 = 16\,\text{mm}$ (M16) (the bolt property class: 8.8) the diameter of bolt circle: $D_2 \geq 2d$ $\xrightarrow{d=80} D_2 \geq 160\,\text{mm}$	No need to calculate it. By referring to the catalog, select a straight teeth-medium duty spline with a diameter of $d \approx 80$ mm and a connection length of $l \geq 80\,\text{mm}$	No need to calculate it. By referring to the catalog, select a rectangular key for the shaft diameter of $d = 80$ mm and a connection length of $l \geq 80\,\text{mm}$	The shaft diameter: $d_1 = 80$ mm

Table 15.6 dimension of the elements attached to the shaft No. 2 of Example 15.2

The dimensions according to the base shaft diameter No. 2 (d = 60 mm) and Table 15.1

Two side weld (under pure torsion) 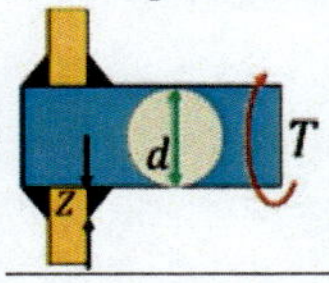	Shaft (pure torsion) 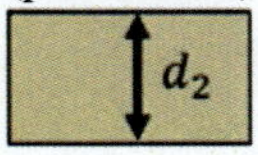	Flange coupling with $z = 8$ 
The width of weld: $Z \geq 0.2d \xrightarrow{d=60} Z \geq 12\,\text{mm}$	The shaft diameter: $d_2 = 60$ mm	The shaft diameter: $d = 60$ mm the bolt size: $d_2 = \frac{d}{5} \xrightarrow{d=60} d_2 = 12\,\text{mm}$(M12) (the bolt property class: 8.8) the diameter of bolt circle: $D_2 \geq 2d \xrightarrow{d=60} D_2 \geq 120\,\text{mm}$

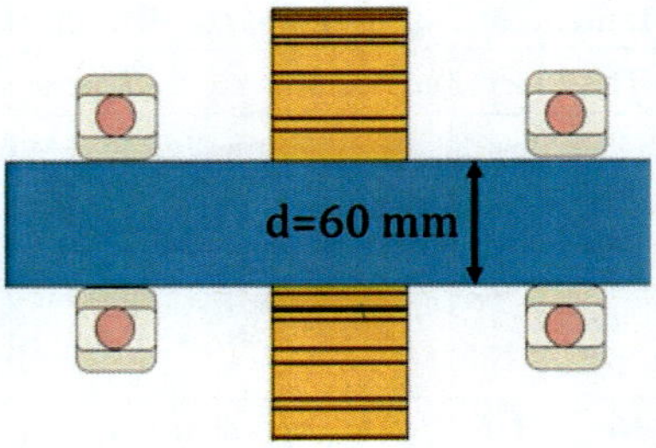

Fig. 15.3 An illustration for the shaft of Example 15.4

Example 15.3 Find the minimum outer diameter of the straight spline of shaft No. 1 in Example 15.1.

Another advantage of M.Y method is that you can estimate dimensions of all the elements using Fig. 15.1.

1. In Fig. 15.1, from the vertical axis of the graph in the second quadrant (the upper left corner) which is related to the splines and rectangular keys, we choose $T = 1500$ N.m and go right to intersect the curve of "Medium-duty".
2. Then, on the curve, we go up to reach the nearest point.
3. From the point, we go vertically down to achieve the minimum outer diameter of the spline shaft, on the horizontal axis:

$$d = 65 \text{ mm}$$

4. Therefore, to transmit a torque of $T = 1500$ N.m, the minimum values for the specification of the straight spline should be:

The outer diameter: $d = 65$ mm.
The engagement length: $l \geq 65$ mm.
And the number of teeth: $n = 8$.

Example 15.4 Consider a shaft with a diameter of $d = 60$ mm, like Fig. 15.3. A gear is supposed to be installed on the shaft, as shown in the figure. Find an appropriate connection for the attachment.

Solution

1. As we have the shaft diameter ($d = 60$ mm), we can easily estimate the attached components using the shaft diameter and Table 15.1. The shaft is similar to the element No. 11 in the table.
2. From Table 15.1, we select the elements that are stronger than the element No. 11, i.e. the elements No. 1 to No.10 (as shown in Table 15.7).
3. As suggested, always consider shafts under bending and torsion. Thus, we can ignore the elements of Nos. 2, 7 and 8, which are under pure torsion.

Table 15.7 the elements that are potentially appropriate for the shaft of Example 15.4

No. 1	No. 2	No. 3	No. 4	No. 5
Press-fit joint	Two side weld (und… pure …rsion)	Involute spline; Conical joint	Straight spline (heavy duty)	Cone ring coupling type A
No. 6	**No. 7**	**No. 8**	**No. 9**	**No. 10**
Flange coupling	Shaft (pure torsion), Ch… …pling	One side weld (under pure …)	Flange coupling	Straight spline (medium duty)

Fig. 15.4 The designed shaft of Example 15.4 with a gear connected to the shaft by a press-fit connection

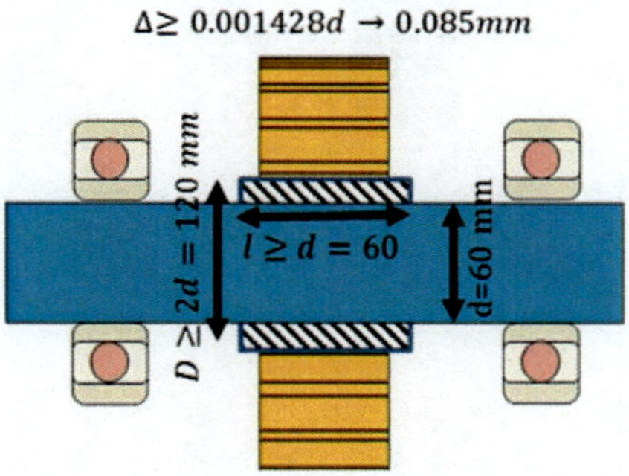

4. Since the aim is to install a gear directly on the shaft, we do not need to use flange couplings. Hence, we can remove the elements of Nos. 6 and 9 from the list too.
5. In the end, we should select one of the rest of the connections.
6. Here, we choose the press-fit connection. The dimensions of the press-fit connection can be obtained using Table 15.1, as shown in Fig. 15.4.

15.3 Equations

Press-fit joint	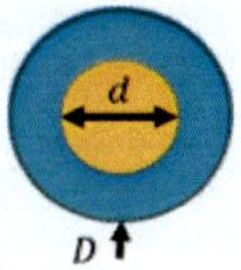	$l = d, D \geq 2d$ $\mu = 0.1$ $\sigma = 300$ MPa $T = 0.0176d^3$ $\Delta = 0.001428d$	Conical joint	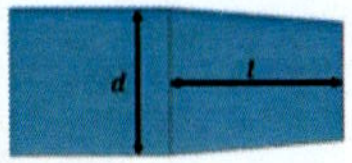	$l = d$ $\mu = 0.1$ $\sigma = 100$ MPa $T = 0.014d^3 n$
One side weld	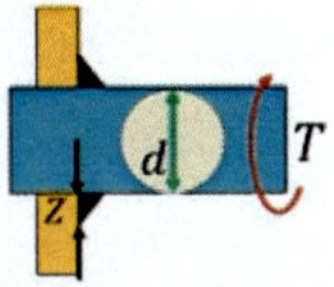	$Z = 0.2d$ $\tau = 35\text{MPa}$ $T_{torsion} = 0.0077d^3$ $T_{torsion\ and\ bending} = 0.0019d^3$	Two side weld	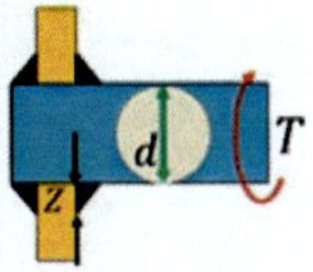	$Z = 0.2d$ $\tau = 35\text{MPa}$ $T_{torsion} = 0.0154d^3$ $T_{torsion\ and\ bending} = 0.0038d^3$
Rectangular key	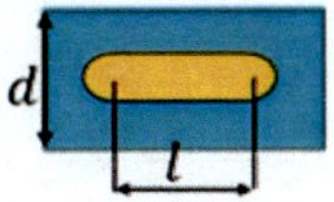	$l = d$ $\sigma = 100$ MPa $T = 0.024d^{2.5414}$	Flange coupling	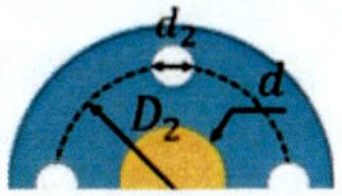	$d_2 = d/5$ $D_2 = 2d$ the bolt property class : 8.8 $T = 0.0012zd^3$ $z = 6, 8, 10$

(continued)

(continued)

Press-fit joint	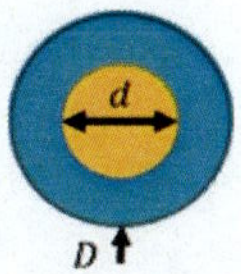	$l = d, D \geq 2d$ $\mu = 0.1$ $\sigma = 300$ MPa $T = 0.0176d^3$ $\Delta = 0.001428d$	Conical joint	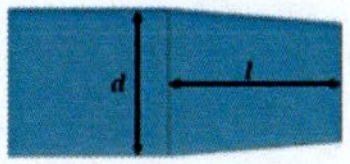	$l = d$ $\mu = 0.1$ $\sigma = 100$ MPa $T = 0.014d^3 n$
Involute spline		$p = 40$ MPa $l = d$ $T = 0.0076d^{3.1364}$	Straight spline		$p = 40$ MPa $l = d$
Cone ring coupling type A		$T = 0.154d^{2.3847}$	Cone ring coupling Type B	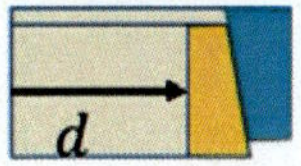	$T = 0.0138d^{2.265}$
Radial pin	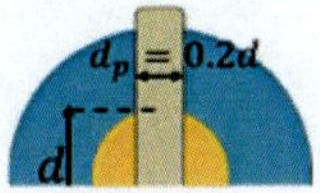	$d_p = 0.2d$ $\tau = 60$MPa $T = 0.00188d^3$	Axial pin	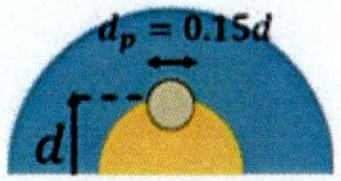	$d_p = 0.15d$ $l = d$ $p = 90$ MPa $T = 0.0033d^3$ Number of pin : 1

Reference

Wittel H, Muhs D, Jannasch D, Voßiek J (2013) Roloff/Matek Maschinenelemente Normung, Berechnung, Gestaltung. Springer Vieweg, Wiesbaden (in German)

Chapter 16
V-Belts

Nomenclature

a	Center distance, mm
D_1	Diameter of small pulley, mm
D_2	Diameter of large pulley, mm
d	Shaft diameter, mm
F	Force required for checking belt tension, N
F_0	Initial tension of each belt, N
f	Belt deflection, mm
L	Belt length, mm
n_1	Rotational speed of small pulley or sheave, rpm
P	Transmitted power, kW
P^*	Specific (or design) power of belt, kW
T_1	Torque on small pulley, N m
u	Speed ratio
v	Belt linear speed, m/s
z	Number of belt

16.1 Introduction

V-belts are flexible elements that are used to transmit power and motion between shafts, which have a relatively long distance from each other. They are perhaps the most widely employed belt drives in machines. V-belts are run in grooved pulleys, or sheaves. The V-shaped cross section of both the V-belt and pulley causes a tight fit.

M. Yaghoubi and H. Tavakoli, *Mechanical Design of Machine Elements by Graphical Methods*, Materials Forming, Machining and Tribology,
https://doi.org/10.1007/978-3-031-04329-1_16

There are several types of V-belts, however, in this chapter we only discuss the normal (or standard) and narrow-section V-belts. Table 16.1 presents some commonly used types and their torque capacity (Zhukov and Gurevich 2014).

This chapter deals with introducing our proposed graphical method for designing V-belts.

16.2 Design Methods

Design of V-belts includes determining type of belt (normal or narrow), diameter of pulleys, center distance and number of belts. In the following subsections, we propose two simple methods for this purpose.

16.2.1 Method I

In this method, you should follow the steps below for designing V-belts (also see Fig. 16.1):

1 First, from Fig. 16.2, choose the type of V-belt (normal or narrow).
2 Then, from Fig. 16.3, find an approximate value for diameter of small pulley (D_1).
3 Then, determine the center distance (a) of the pulleys using Fig. 16.3.
 Note: The acceptable range of center distance is calculated from Eq. 16.1 (see Sect. 16.4) (Jiang 2019):
4 Then, using Fig. 16.4, check whether the rotational speed of pulley is suitable or not.
5 Finally, specify the number of belts (z) by employing Figs. 16.5, 16.6, 16.7 and Eq. 16.2 (see Sect. 16.4) ($z \leq 9$).

Example 16.1 A V-belt is supposed to transmit a power of $P = 160$ kW, between two shafts as shown in Fig. 16.1. Design the V-belt for this task.

Solution

To design the V-belt, we need to specify the type of belt, the diameter of small pulley (D_1), the diameter of large pulley (D_2), the center distance (a) and the number of belts (z).

1 First from Fig. 16.2, we choose the type of belt (normal or narrow). Because the size of pulleys for the narrow belts is smaller, we select the narrow type arbitrarily.
2 Therefore, using the right side graph of Fig. 16.2, we find $P = 160$ kW on the horizontal axis and go up. At the same time, we choose $n = 1000$ rpm on the

Table 16.1 Types of V-belts and their torque capacity range

Type of V-belt	Normal V-belts						Narrow-section V-belts			
	Z	*A*	*B*	*C*	*D*	*E*	*SPZ*	*SPA*	*SPB*	*SPC*
Torque capacity (N m)	≤ 63	11–70	40–196	110–550	450–2000	1100–4500	≤ 150	90–400	300–2000	2000–8000

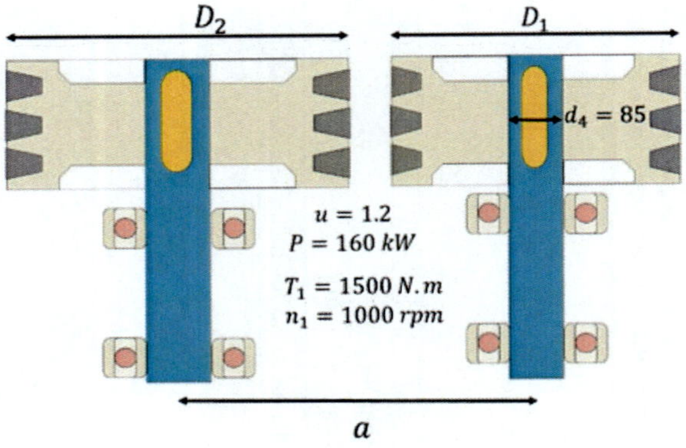

Fig. 16.1 An illustration for Example 16.1

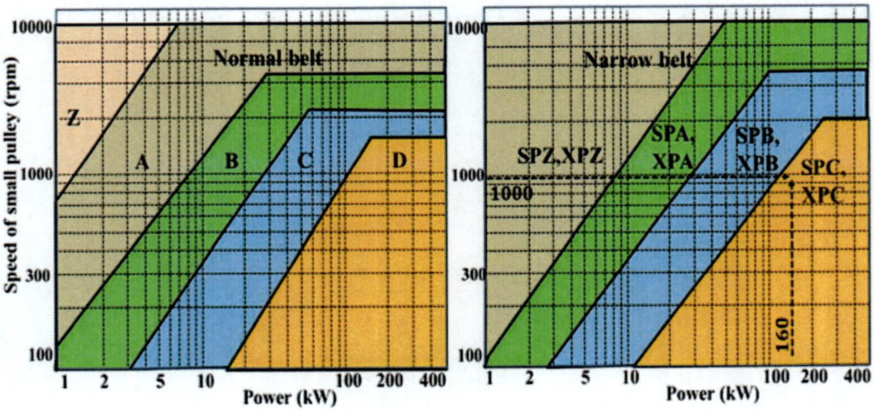

Fig. 16.2 Graphs for selecting the V-belt type based on the rotational speed of small pulley, and power (plotted based on data of SKF Group 2016)

vertical axis and go right. The intersection point is in the area of the types SPC and XPC.

3 We can select either of them. The XPC type is a little stronger and more expensive than the SPC type. However, we prefer to choose the XPC type.

4 From the right side graph of Fig. 16.3, we can find the approximate diameter of small pulley (D_1).

5 On the horizontal axis, we choose $T_1 = 1500$ N m, and go up to cross either of the two curves related to the narrow belt type, "GOST" (Feshenko 2016) or "Zhukov" (Zhukov and Gurevich 2014). Here, we choose the curve "GOST". Then, we go left to find the diameter:

$$D_1 \approx 340 \text{ mm}$$

(If we chose the curve of "Zhukov", the diameter would be $D_1 \approx 230$ mm.)

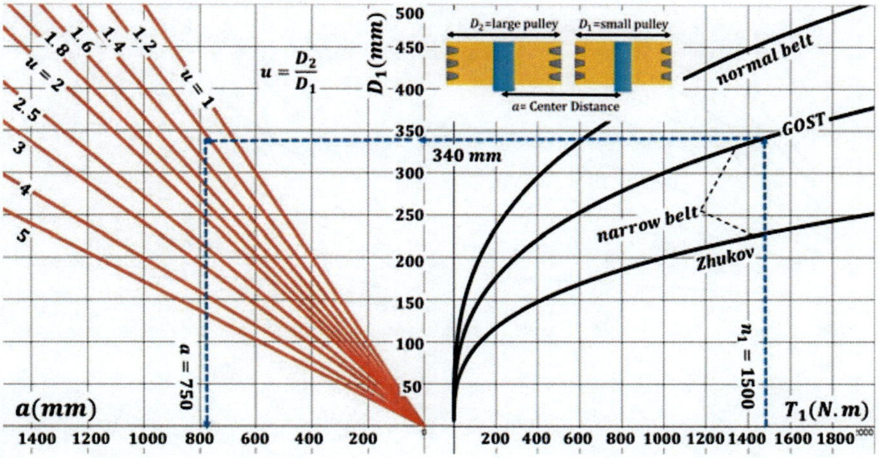

Fig. 16.3 Graphs for determining the diameter of small pulley (D_1) and the center distance (a) for V-belts

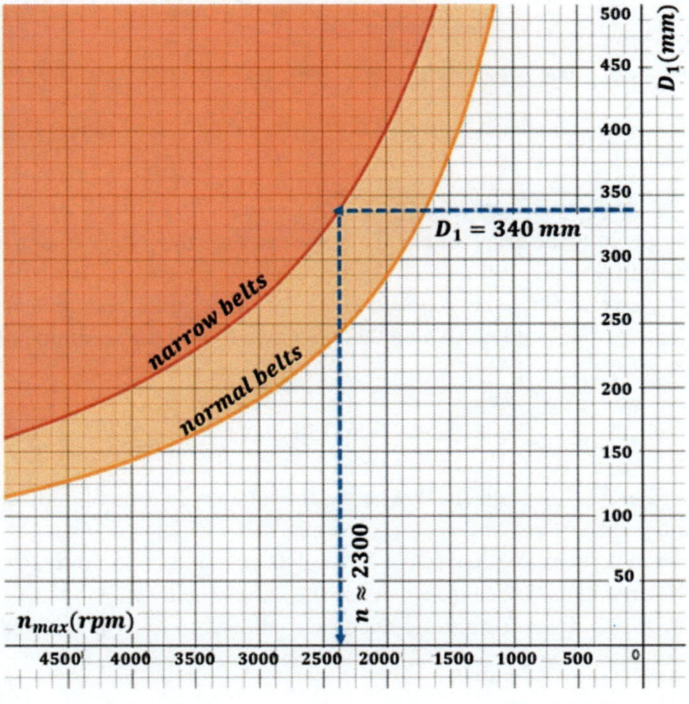

Fig. 16.4 A graph representing the relationship between the diameter (D_1) and the maximum allowable rotational speed (n_{max}) of small pulley

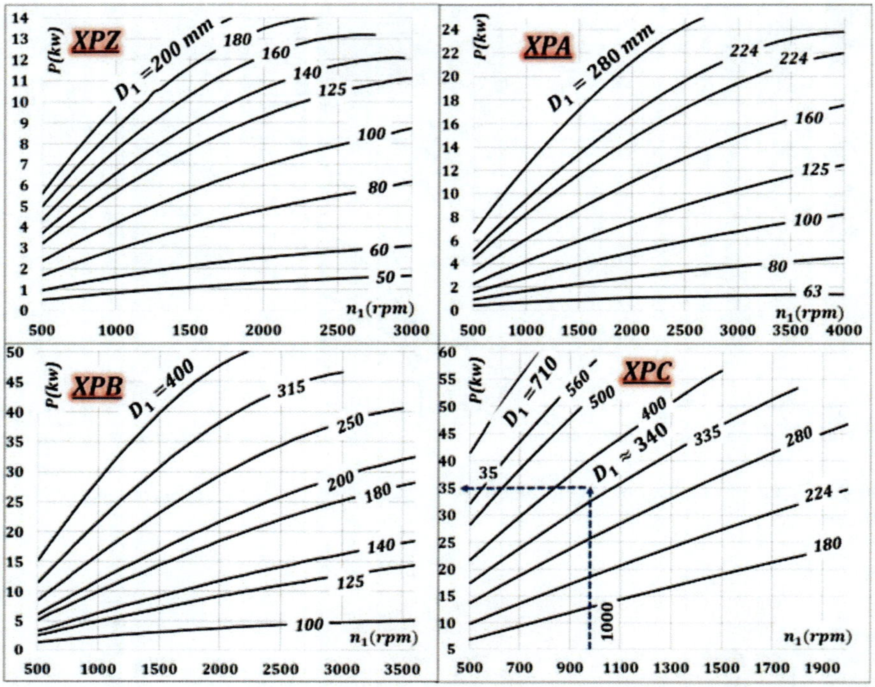

Fig. 16.5 Specific power of the V-belts XPZ, XPA, XPB and XPC as a function of the rotational speed of pulley

6 Using the left side graph of Fig. 16.3, we can determine the center distance (a) of the pulleys.

7 From the vertical axis, we choose $D_1 \approx 340$ mm, go left to intersect the curve of $u = 1.2$ (the speed ratio), and then go down to find the recommended center distance:

$$a \approx 750 \text{ mm}$$

8 Using Fig. 16.4, we can check whether the rotational speed of pulley is suitable or not.

9 From the vertical axis, we select $D_1 \approx 340$ mm, and go left to reach the curve of "Narrow belts" (because we chose the narrow belt XPC).

10 Then, we go down to find the maximum allowable rotational speed of small pulley:

$$n_{max} = 2400 \text{ rpm}$$

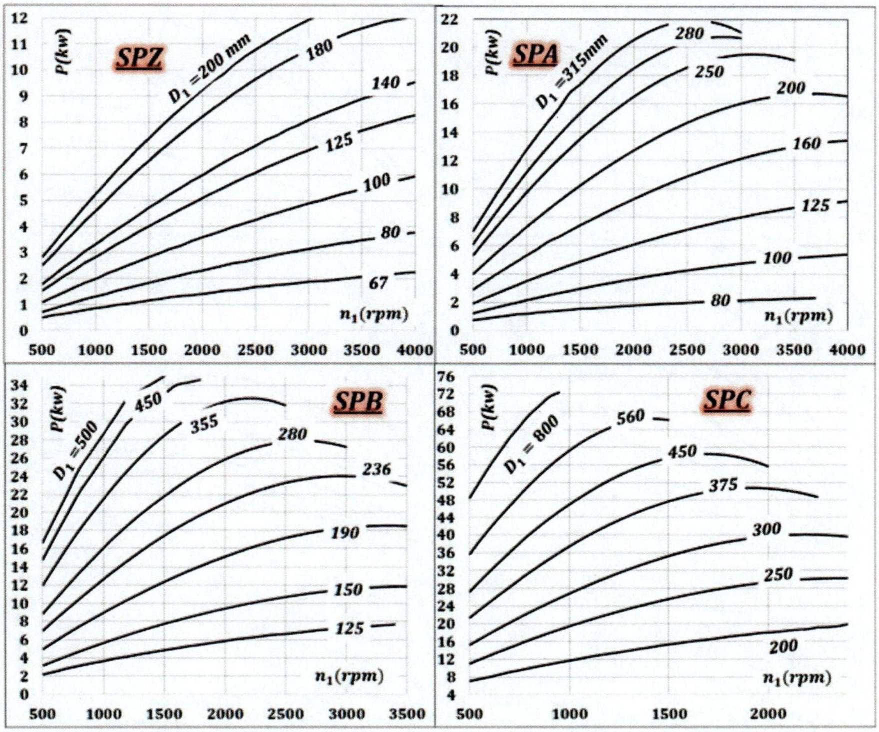

Fig. 16.6 Specific power of the V-belts SPZ, SPA, SPB and SPC as a function of the rotational speed of pulley

The rotational speed of the small pulley in this example is $n = 1000$ rpm which is smaller than $n_{max} = 2400$ rpm. Thus, the diameter of small pulley ($D_1 \approx 340$ mm) has been selected correctly.

11 Finally, we should determine the number of belts (z) using Fig. 16.5 and Eq. 16.2 (see Sect. 16.4).

12 For this purpose, first we should find the value of specific power (P^*) from Fig. 16.5.

The graph in the fourth quadrant (the lower right corner) is related to the XPC belt type. On the horizontal axis, we select $n_1 = 1000$ rpm, go up to cross the approximate curve of $D_1 = 340$ mm, and then go left to find the specific power:

$$P^* = 35 \text{ kW}$$

13 Now, we can calculate the number of belts using Eq. 16.2 (see Sect. 16.4):

$$z \approx 1.3\frac{P}{P^*} \xrightarrow{P^*=35 \text{ kW, and } P=160 \text{ kW}} z = 1.3 \times \frac{160}{35} = 5.9 \rightarrow z \geq 6$$

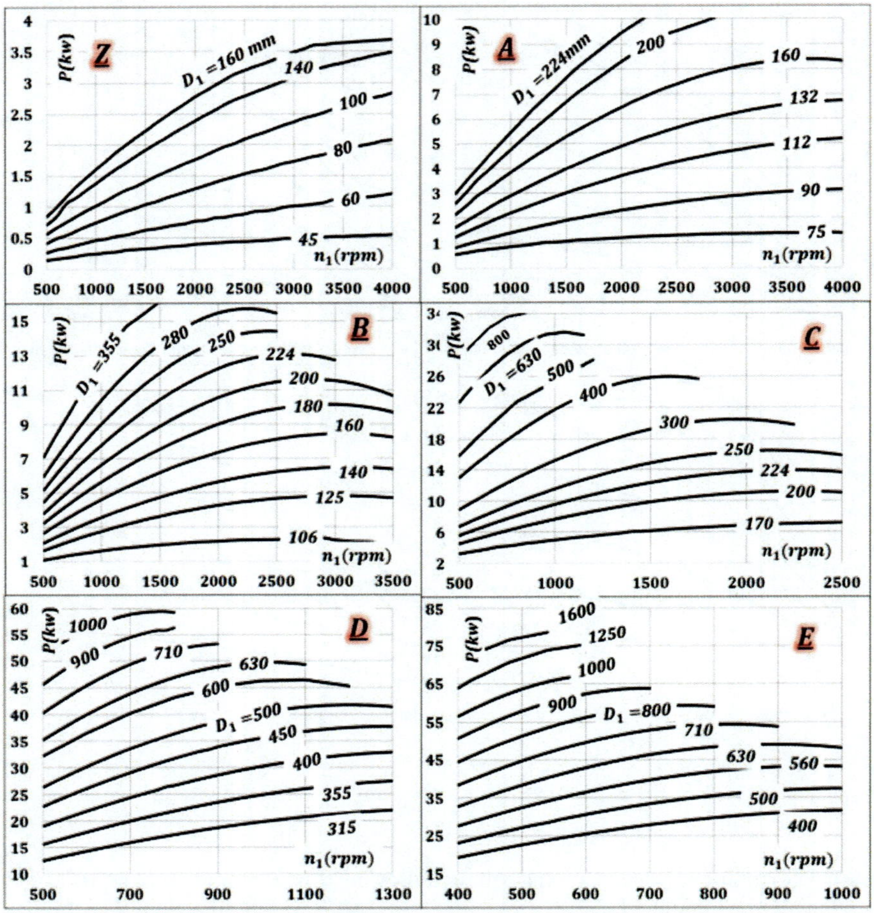

Fig. 16.7 specific power of the V-belts Z, A, B, C, D and E as a function of the rotational speed of pulley

14 The number of belts obtained $z = 5.9$ and we round it up to 6. However, for safety reasons, we select 7 belts for the V-belt of this example ($z = 7$).

15 The characterizations of the V-belt is summarized below:

Type of belt: XPC

Diameter of the small pulley: $D_1 = 340$ mm

Diameter of the large pulley: $D_2 = u \times D_1 = 1.2 \times 340 = 408$ mm

Center distance: $a = 750$ mm

Number of belts: $z = 7$

The belt length can be calculated using Eq. 16.3 (see Sect. 16.4) (Jiang 2019):

Table 16.2 Relationships between the diameter of small pulley (D_I) and the shaft diameter (d) according to M.Y method

Shaft arrangement	Symmetrical or asymmetrical 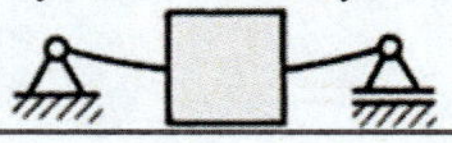	Cantilever 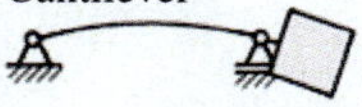
Narrow belts	$D_1 = (4 - 5.4)d$	$D_1 = (3.5 - 4.3)d$
Normal belts	$D_1 = 7.2d$	$D_1 = 5.7d$

$$\mathrm{L} = 2a + \frac{\pi}{2} D_1 \times (1 + u) + \frac{D_I^2 \times (u - 1)^2}{4a}$$
$$\approx 2a + 1.57 \times D_1 \times (1 + u) \rightarrow L = 2674 \text{ mm}$$

Note: If you choose the center distance (a) from Fig. 16.3, the approximate value of the belt length (L) can be calculated using Eq. 16.4 (see Sect. 16.4). Therefore, we have: $\mathrm{L} \approx 3.58 D_1 \times (1 + u) \rightarrow L = 2677$ mm. Afterwards, you should round it up to the standard lengths according to manufacturers' catalogs.

16.2.2 Method II (M.Y Method)

In M.Y method, if the shaft diameter (d) is known, the diameter of small pulley (D_I) can be estimated according to the shaft diameter, as shown in Table 16.2.

Example 16.2 In Example 16.1, suppose that the shaft diameter for the small pulley is $d = 85$ mm and the shaft arrangement is cantilever. Determine the diameter of small pulley?

Solution

We choose the narrow belt type. As the shaft arrangement is cantilever, we have:

$$D_I = (3.5 - 4.3)d = (3.5 - 4.3) \times 85 = 297.5 - 365.5 \text{ mm}$$

16.3 Belt Tensioning

One of the methods for checking the belt tension is illustrated in Fig. 16.8. In this method, a force (F) is applied to the middle of each belt and the resultant deflection (f) is measured. The deflection can be estimated using Eq. 16.5 (see Sect. 16.4) and the amount of force to be applied for the check, can be found using Fig. 16.9. The following example shows how we apply this method.

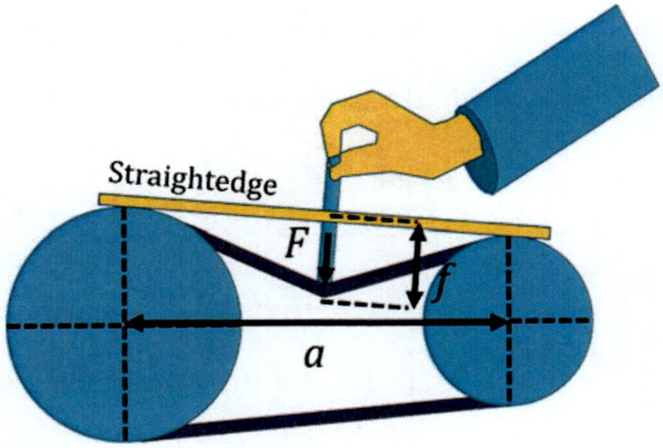

Fig. 16.8 An illustration for checking the belt tension

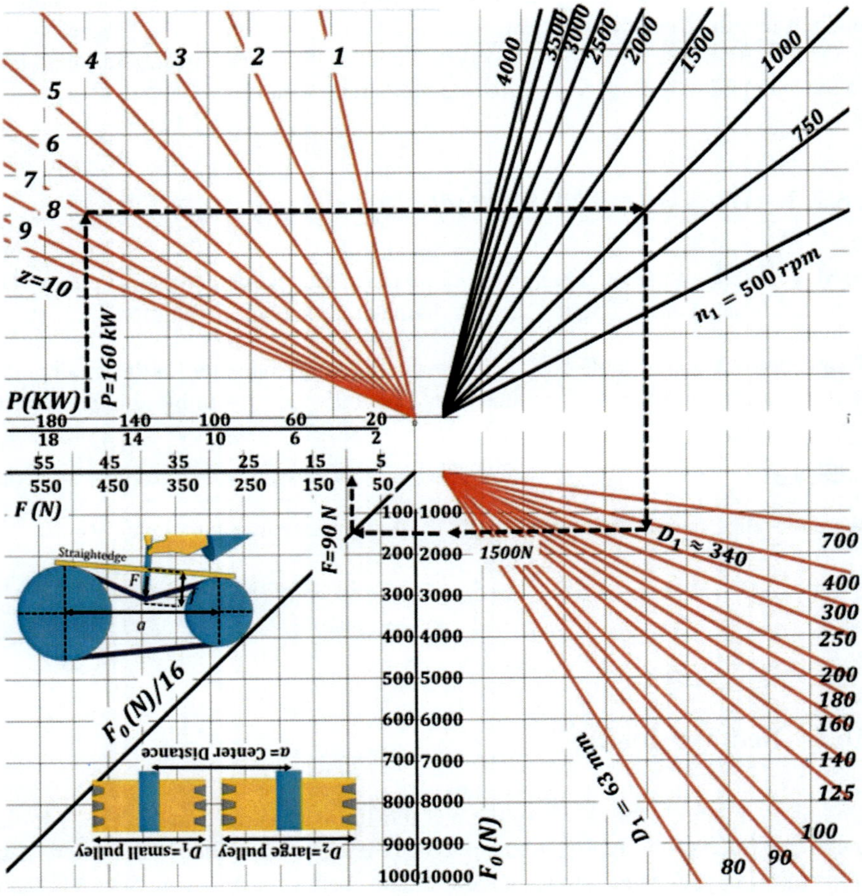

Fig. 16.9 Graphs for determining the initial tension of belt, and the force required for checking the belt tension

Example 16.3 Determine the belt tensioning force and deflection for the V-belt of Example 16.1.

Solution

The obtained values for the parameters of the V-belt in Example 16.1 are:

$$P = 160 \text{ kW}, z = 7, n_l = 1000 \text{ rpm}, a = 750 \text{ mm}, \text{ and } D_l = 340 \text{ mm}$$

1. First, in the second quadrant (the upper left corner) of Fig. 16.9, we select $P = 160$ kW on the horizontal axis, and go up to reach the curve related to $z = 7$ (number of belts).
2. Then, we go right to cross the curve of $n_l = 1000$ rpm (rotational speed of small pulley).
3. Then, we go down to intersect the approximate curve of $D_l = 340$ mm (diameter of small pulley).
4. After that, we go left to reach the vertical axis which yields the initial tension of belt:

$$F_0 = 1500 \text{ N}$$

(The initial tension (F_O) can be calculated using Eq. 16.6 (see Sect. 16.4) too.)

5. We keep going left to cross the curve of "$\frac{F_0}{16}$".
6. Then, we go up to find the force on the horizontal axis:

$$F = 90 \text{ N}$$

7. In this stage, we estimate the belt deflection (f) using Eq. 16.5 (see Sect. 16.4):

$$f \approx \frac{a}{63} \approx \frac{750}{63} \rightarrow f \approx 11 \text{ mm}$$

In conclusion, the deflection of each sbelt after applying the force of $F = 90$ N should be $f \approx 11$ mm.

16.4 Equations

$$0.7(D_1 + D_2) \leq a \leq 2(D_1 + D_2) \tag{16.1}$$

$$z \approx 1.3\frac{P}{P*} \tag{16.2}$$

$$\mathrm{L} = 2a + \frac{\pi}{2}D_l(1 + u) + \frac{D_l^2(u-1)^2}{4a} \tag{16.3}$$

$$L \approx 3.58D_1(1+u) \tag{16.4}$$

$$f \approx \frac{a}{63} \tag{16.5}$$

$$F_0 \approx 500 \times \frac{2.5 - c_\propto}{c_\propto} \times \frac{1.3P}{znD_1} \times 19100$$
$$c_\propto = 0.9 \tag{16.6}$$

The equations of Fig. 16.3:

$$D_1 = C\sqrt[3]{T_1}$$
$$C = \begin{cases} 40 \text{ (for normal V-belts)} \\ 20 \text{ (for narrow V-belts-Zhukov)} \\ 30 \text{ (for narrow V-belts-Ghost)} \end{cases}$$
$$a = 2D_1\sqrt[3]{u^2} \tag{16.7}$$

The equation of Fig. 16.4:

$$\frac{n_1D_1}{19100} \leq V$$
$$v = \begin{cases} 30 \text{ (for normal V-belts)} \\ 42 \text{ (for narrow V-belts)} \end{cases} \tag{16.8}$$

References

Feshenko VN (2016) Designer's handbook. Book 1. Machines and mechanisms: textbook-practical. Infra-Engineering, Moscow-Vologda (in Russian)

Jiang W (2019) Analysis and design of machine elements, 1st edn. Wiley, New Jersey

SKF Group (2016) Retrieved from https://www.skf.com/group

Zhukov KP, Gurevich YE (2014) Design of parts and assemblies of machines: a textbook for universities. Mechanical Engineering, Moscow (in Russian)

Chapter 17
Timing Belts

Nomenclature

a	Center distance, mm
b	Belt width, cm or mm
D_1	Diameter of small pulley, mm
D_2	Diameter of large pulley, mm
F	Force required for checking belt tension, kg
F_t	Peripheral force on the belt, N
F_{zul}	Permissible force on the belt, kg or N
f	Belt deflection, mm
n_1	Rotational speed of small pulley, rpm
P	Transmitted power, kW
P^*	Specific (or design) power of the belt, kW
p	Pitch, mm
T	Torque, N m
u	Speed ratio
z_1	Teeth number of small pulley
z_2	Teeth number of large pulley
β	Angle of belt wrap, degrees

17.1 Introduction

Timing belts, also known as synchronous belts or toothed belts, perform power and motion transmissions by means of teeth that are on the underside of the belt. This is in contrast to other types of belts in which the transmission is done by friction. Therefore, in timing belts, almost no slip or creep occurs and the driving and driven

M. Yaghoubi and H. Tavakoli, *Mechanical Design of Machine Elements by Graphical Methods*, Materials Forming, Machining and Tribology,
https://doi.org/10.1007/978-3-031-04329-1_17

Fig. 17.1 Cross section view of different types of timing belts

shafts are kept synchronized. This makes timing belts suitable for many applications such as driving an engine camshaft by the crankshaft, in which the synchronization is essential. Toothed wheels or pulleys run timing belts (Juvinall and Marshek 2011; Mott et al. 2018; Jiang 2019).

Timing belts with different tooth profile shapes are available as shown in Fig. 17.1. In this book, we only discuss design of trapezoid and High Torque Drive (HTD) types. The trapezoid type of timing belts are usually categorized into T2.5, T5, T10, and T20 types, in which the number indicates the size of tooth pitch in mm. The HTD type has different classes of 3 M, 5 M, 8 M, 14 M, and 20 M, in which the number stands for the size of tooth pitch in mm.

17.2 Design Methods

17.2.1 Initial Estimate

Design of timing belts includes determining the size of tooth pitch (in other words the belt type) and belt width. Equations 17.1 and 17.2 can be used for this purpose (see Sect. 17.3).

However, for simplicity, rather than using the equations, we propose a graphical method for the design. The design process includes:

1 First, specifying the type of belt to be used, trapezoid type or HTD type.
2 Then, determining the tooth pitch of the belt from Fig. 17.2, according to the rotational speed of small pulley (n_1) and the transmitted power (P).
3 Estimating the belt width using the graphs of Figs. 17.3 or 17.4.
4 In the end, calculating other parameters from Eqs. 17.3, 17.4 and 17.5 (see Sect. 17.3).

The following examples illustrate applying the method.

Example 17.1 A power of $P = 160$ kW is supposed to be transmitted between two shafts by means of a timing belt as shown in Fig. 17.5. The specifications of the power transmission system is given in Fig. 17.5a. Determine the belt type, tooth pitch, teeth number of small and large pulleys, and diameter of small and large pulleys.

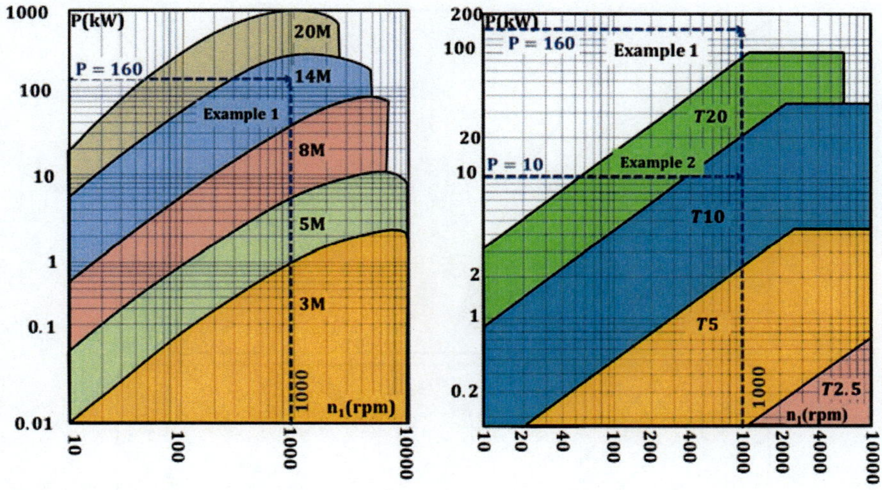

Fig. 17.2 Graphs for specifying the type of timing belt

Solution

1 First, we determine the belt type using Fig. 17.2.

2 On the both sides of Fig. 17.2, we select $P = 160$ kW from the vertical axis and go right. At the same time, we select $n_1 = 1000$ rpm from the horizontal axis of the both sides, and go up. The intersections on the left and right diagrams, specifies the belt type for HTD and trapezoid, respectively. As it can be seen, for T-type belts, the intersection point is above T20 that means the trapezoid belts are weak for transmitting this power. However, for HTD type belts, the point of intersection is in the area of 14 M. Thus, we select 14 M belt type.

3 In Fig. 17.3d, which is related to 14 M belts, we select $n = 1000$ rpm from the left horizontal axis and go up to cross the curve of $D_1 = 214$ mm. Here, we arbitrarily choose 214 mm for the diameter of small pulley.

Then, we go right.

4 At the same time, from the right horizontal axis of Fig. 17.3d, we select $P =$ 160 kW and go up. The intersection indicates the belt width:

$$b \approx 200 \text{ mm}$$

In the end, the specifications of the power transmission system are as follows (Fig. 17.5b):

The timing belt type: 14 M.

The tooth pitch of the belt: $p = 14$ mm.

The belt width: $b \geq 200$ mm

The teeth number of small pulley, using Eq. 17.3 (see Sect. 17.3):

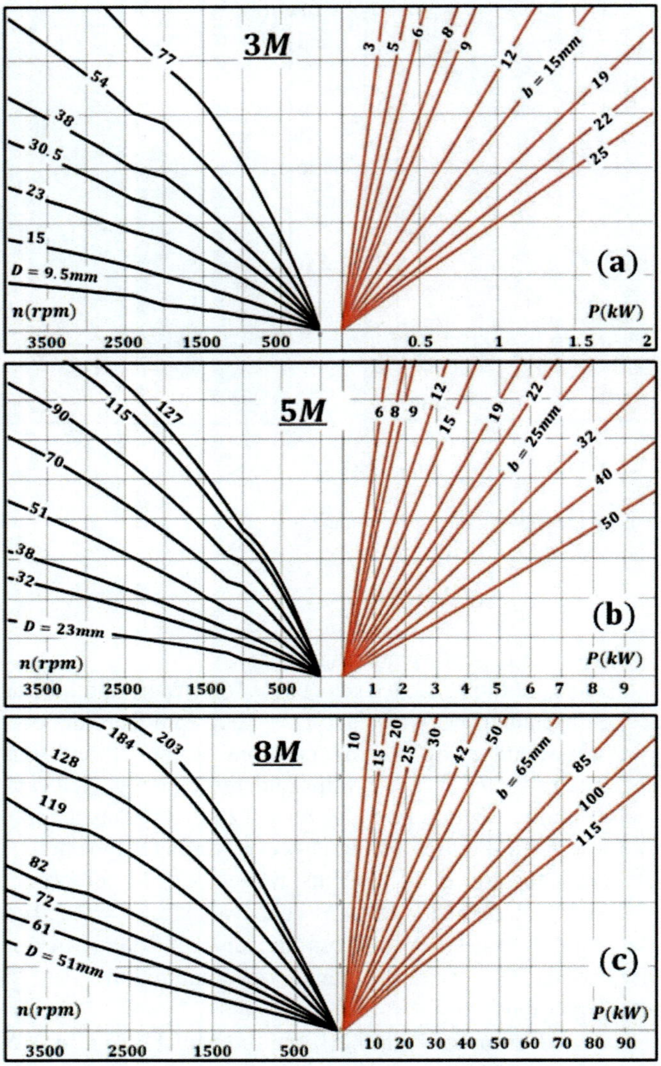

Fig. 17.3 Graphs for estimating the width of different HTD belt types, **a** the type 3 M, **b** the type 5 M, **c** the type 8 M, **d** the type 14 M, and **e** the type 20 M

$$z_1 = \frac{D_1 \times \pi}{p} = \frac{214 \times \pi}{14} \approx 48 \text{ mm}$$

The teeth number of large pulley, using Eq. 17.4 (see Sect. 17.3):

$$z_2 = u \times z_1 = 1.2 \times 48 = 57.6 \approx 58$$

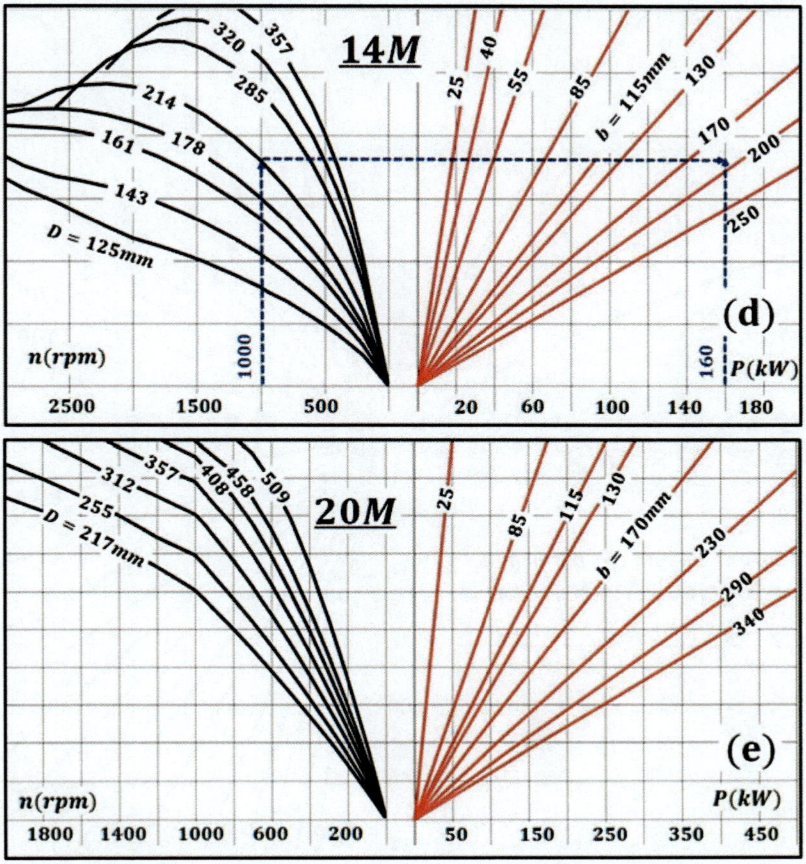

Fig. 17.3 (continued)

The diameter of small pulley: $D_1 = 214$ mm
The diameter of large pulley, using Eq. 17.4 (see Sect. 17.3):

$$D_2 = u \times D_1 = 1.2 \times 214 \approx 256.8 \text{ mm}$$

The center distance, using Eq. 17.5 (see Sect. 17.3):

$$0.5(D_1 + D_2) + 15 \leq a \leq 2(D_1 + D_2) \rightarrow 250.4 \text{ mm} \leq a \leq 941.6 \text{ mm}$$

Example 17.2 Suppose that in Example 17.1, a power of $P = 10$ kW is transmitted, rather than $P = 160$ kW. Select a T-type belt for it and estimate the belt width?

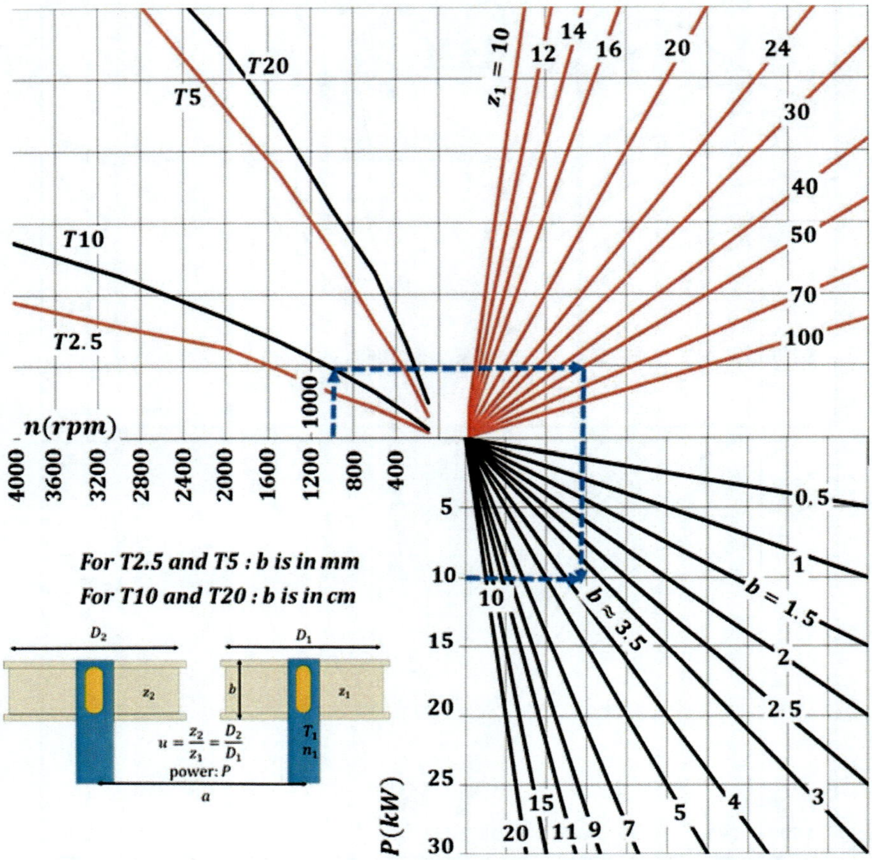

Fig. 17.4 Graphs for estimating the width of T-type (trapezoid) belts

Solution

1 First, from the right diagram of Fig. 17.2, we find $P = 10$ kW on the vertical axis and go right. Simultaneously, from the horizontal axis of the same diagram, we select the rotational speed of small pulley, $n_1 = 1000$ rpm, and go up. The intersection point is in the area related to T10. Therefore, the belt type is T10.

2 Now, in Fig. 17.4, we select $n_1 = 1000$ rpm from the left horizontal axis and go up to reach the curve of T10.

3 Then, we go right to cross the curve related to $z_1 = 50$. We arbitrarily select 50 for the teeth number of small pulley. However, you can use Fig. 17.6 to determine the suggested minimum number of teeth as a function of the speed ratio.

After that, we go down.

4 At the same time, we select $P = 10$ kW from the lower vertical axis and go to right. The point of intersection determines the size of the belt width:

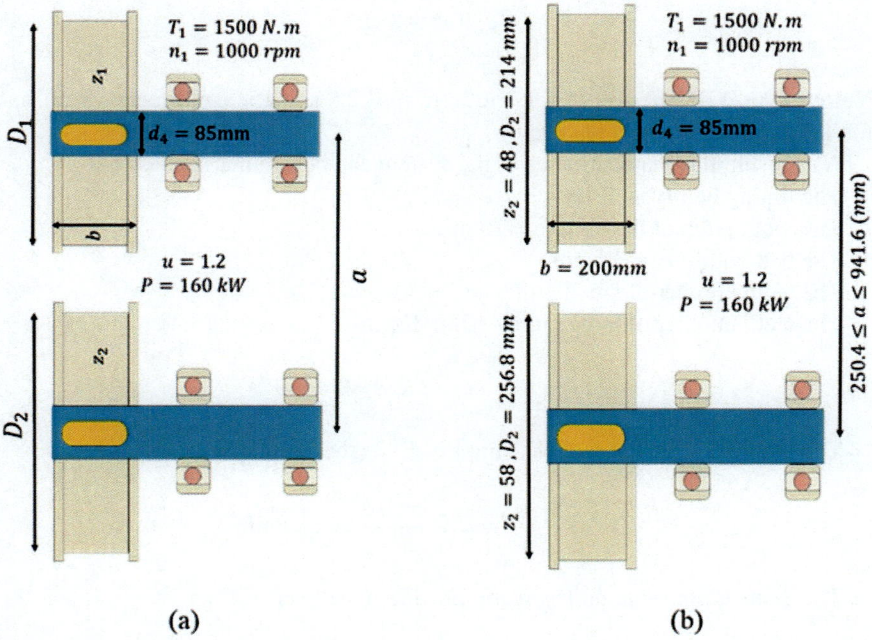

Fig. 17.5 An illustration for the power transmission system of Example 17.1

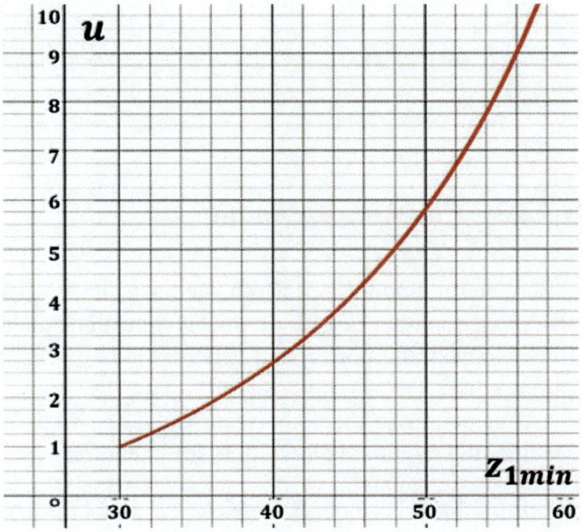

Fig. 17.6 A graph for determining the suggested minimum number of teeth of small pulley (z_1) as a function of the speed ratio (u)

$$b \approx 1.8 \text{ cm}$$

Note: In Fig. 17.4, the unit of belt width (b) for T2.5 and T5 is millimeter (mm), and for T10 and T20 is centimeter (cm).

As a result, the characteristics of the system can be summarized as follows:
The timing belt type: T10.
The tooth pitch of the belt: $p = 10$ mm.
The belt width: $b \geq 3.5$ cm
The teeth number of small pulley: $z_1 = 50$
The teeth number of large pulley, using Eq. 17.4 (see Sect. 17.3):

$$z_2 = u \times z_1 = 1.2 \times 50 = 60$$

The diameter of small pulley, using Eq. 17.3 (see Sect. 17.3):

$$D_1 = \frac{z_1 \times p}{\pi} = \frac{50 \times 10}{\pi} = 160 \text{ mm}$$

The diameter of large pulley, using Eq. 17.4 (see Sect. 17.3):

$$D_2 = u \times D_1 = 1.2 \times 160 = 192 \text{ mm}$$

The center distance, using Eq. 17.5 (see Sect. 17.3):

$$0.5(D_1 + D_2) + 15 \leq a \leq 2(D_1 + D_2) \rightarrow 191 \text{ mm} \leq a \leq 704 \text{ mm}$$

17.2.2 Checking Estimations

After determining the belt width, we must check whether the designed belt tolerates the applied load or not. To do this, we must calculate the force, which corresponds to the transmitted power, and acts on the belt. Equation 17.6 can be used for this purpose (see Sect. 17.3).

After obtaining the force from Eq. 17.6, we can check from Fig. 17.7 to see whether the required belt width is sufficient for withstanding the force or not.

Example 17.3 For both Examples 17.1 and 17.2, check whether the estimated belt width is appropriate or not.

Solution

1 For Example 17.1, the designed belt has the specifications below:
Belt type 14M, $P = 160$ kW, $n_1 = 1000$ rpm, $b = 200$ mm and $D_1 = 214$ mm

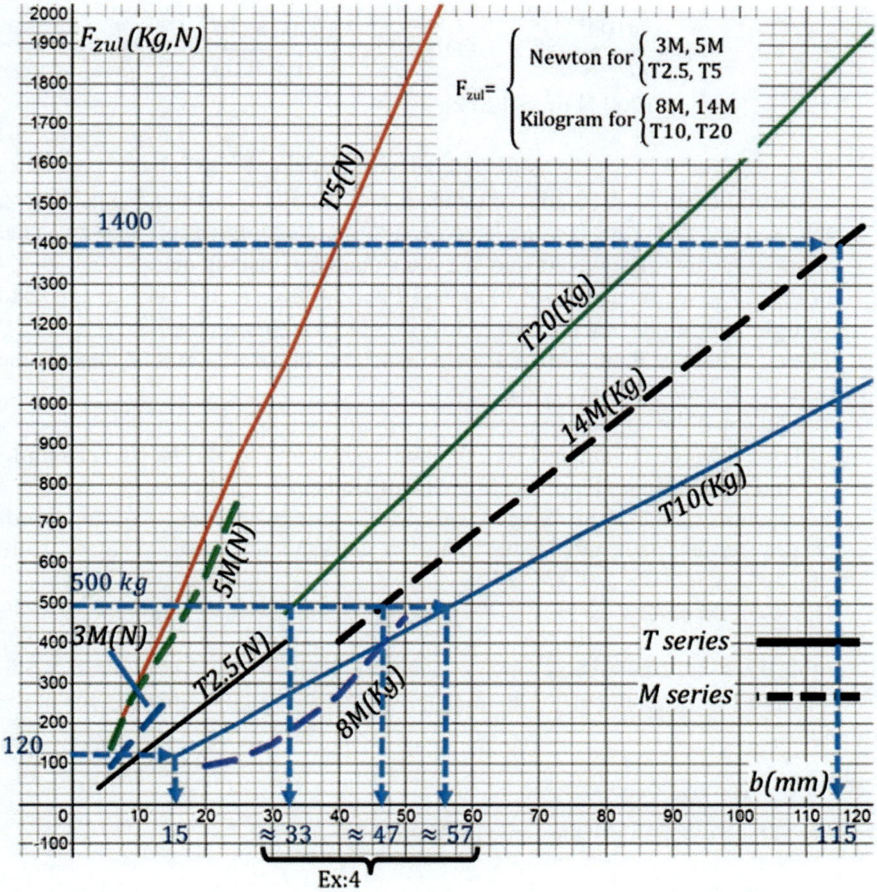

Fig. 17.7 A graph for checking the estimated width of timing belts

We can calculate the peripheral force (F_t) for Example 17.1 using Eq. 17.6 (see Sect. 17.3):

$$F_t = \frac{19100 \times P}{D_1 \times n_1} \times 1000 \rightarrow F_t = \frac{19100 \times 160}{214 \times 1000} \times 1000$$
$$= 14000 \text{ N or} \approx 1400 \text{ kg}$$

2 The specifications of the selected belt for Example 17.2 are:

Belt type T10, $P = 10$ kW, $n_1 = 1000$ rpm, $b = 18$ mm and $D_1 = 160$ mm

For Example 17.2 too, the peripheral force (F_t) can be calculated by applying Eq. 17.6 (see Sect. 17.3):

$$F_t = \frac{19100 \times P}{D_1 \times n_1} \times 1000 \rightarrow F_t = \frac{19100 \times 10}{160 \times 1000} \times 1000$$
$$= 1200 \text{ N or} \approx 120 \text{ kg}$$

3 Now, we should check the estimated belt widths using Fig. 17.7.
Note: In Fig. 17.7, the unit of force (F_t) for T2.5, T5, 3M and 5M is Newton (N), and for T10, T20, 10M and 14M is kilogram (kg). Equation 17.6 gives the value of F_t in N. To convert it to kg, you can approximately divide it by 10.
4 In Fig. 17.7, we select $F_t = 120$ kg and $F_t = 1400$ kg on the vertical axis, and go right to cross the curves related to T10 and 14 M, respectively.
5 Then, for both of them, we go down to reach the horizontal axis and obtain the permissible widths.
6 To withstand the applied forces:
For 14M type (Example 17.1), the minimum width should be $b = 115$ mm.
And for T10 type (Example 17.2), it should be $b = 15$ mm.
7 In conclusion, both the estimated widths in Examples 17.1 and 17.2 are appropriate. Because they are greater than the corresponding ones obtained from Fig. 17.7.
8 However, for safety, we choose $b = 20$ mm for the belt of Example 17.2.

Example 17.4 For a conveyor which is designed using a timing belt (as shown in Fig. 17.8), the force on the belt is about 250 kg. Determine the width and type of the timing belt.

Solution

1 For safety, we multiply the force by 2:

$$F = 250 \times 2 = 500 \text{ kg}$$

2 In Fig. 17.7, we select the force of $F_{zul} = 500$ kg from the vertical axis and go right to cross the curves related to the belt types T20, 14 M, and T10, respectively (because the forces in these four curves are in kg).
3 Then, we go down from all the crossed curves, to obtain the width on the horizontal axis:

Fig. 17.8 Schematic of the conveyor of Example 17.4

For T20: $b = 33$ mm
For 14M: $b = 47$ mm
For T10: $b = 57$ mm

4 We choose the belt type T20 with a width of $b = 33$ mm.

17.2.3 Belt Installation

After designing or selecting timing belts, they are installed on pulleys and tightened. The tightening should be done so that the belt's tension reaches an appropriate amount.

One of the methods for checking the belt tension is shown in Fig. 17.9. In this method, a force (F) is applied to the middle of the belt and the resulted deflection (f) is measured.

To find the values of the force (F) and the deflection (f), you can use Eqs. 17.7 and 17.8 (see Sect. 17.3), respectively, or refer to Fig. 17.10. The following example shows how we apply graphs of Fig. 17.10 for this purpose.

Example 17.5 Determine the amount of the force required for checking the belt tension (F) and the corresponding deflection (f) of the belt in Example 17.1. Assume that the value of center distance is $a = 500$ mm.

Solution

1 From the left horizontal axis of Fig. 17.10, we select the power of $P = 160$ kW, go up to reach the curve of $n_1 = 1000$ rpm, then go right to cross the approximate curve of $D_1 = 214$ mm, and then go down to obtain the force value:

$$F = 38 \text{ kg}$$

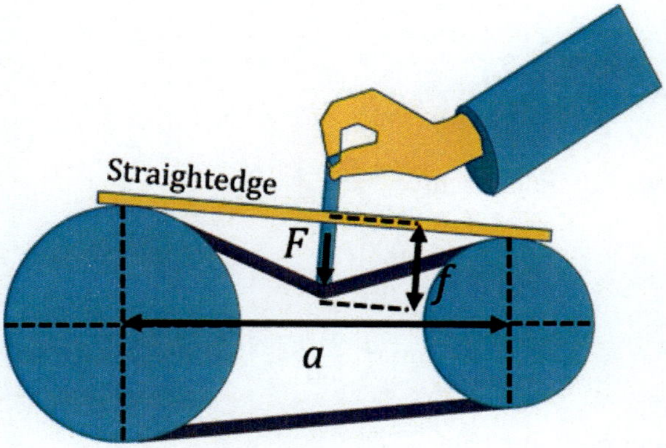

Fig. 17.9 An illustration for checking the tension of timing belts

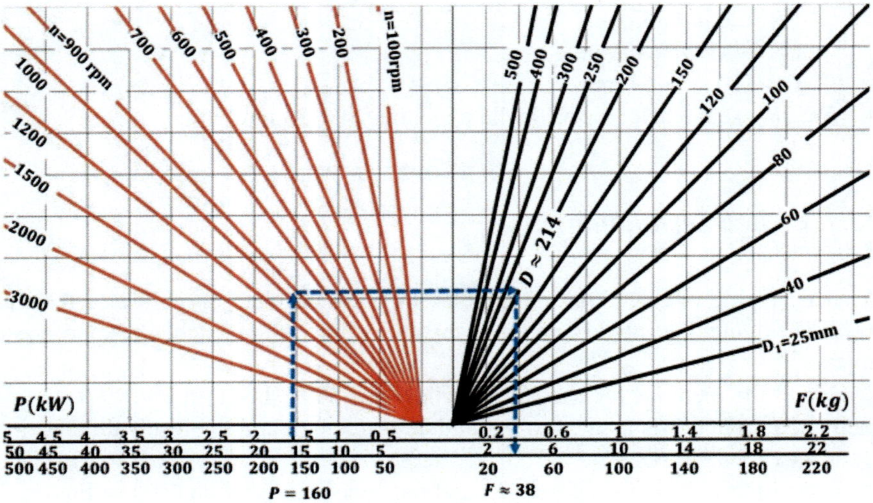

Fig. 17.10 Graphs for checking timing belt tension

2 To determine the amount of deflection (f), we divide the value of $a = 500$ mm by 50 (Eq. 17.8 (see Sect. 17.3)):

$$f = \frac{a}{50} = \frac{500}{50} = 10 \text{ mm}$$

3 In conclusion, the belt is sufficiently tightened if after applying a force of $F = 38$ kg onto the middle of the belt, a deflection of $f = 10$ mm occurs.

Note: If shock and vibration forces are applied to timing belts, multiply the value of F obtained from Fig. 17.10 by 2.

17.3 Equations

For trapezoid types of timing belts:

$$b \geq \frac{1000P}{P^{*}k}$$

$$k = \min\left(\frac{\beta}{360}z_1^2, 12z_1\right) \tag{17.1}$$

For HTD types of timing belts:

$$1000P = P^{*}b \tag{17.2}$$

$$z_1 = \frac{\pi D_1}{p} \tag{17.3}$$

$$u = \frac{z_2}{z_1} = \frac{D_2}{D_1} \tag{17.4}$$

$$0.5(D_1 + D_2) + 15 \leq a \leq 2(D_1 + D_2) \tag{17.5}$$

$$F_t = \frac{19100P}{D_1 n_1} \times 1000 \tag{17.6}$$

$$F = \frac{19100P}{D_1 n_1} \times (2.5 - 5) \tag{17.7}$$

$$f \approx \frac{a}{50} \tag{17.8}$$

References

Jiang W (2019) Analysis and design of machine elements, 1st edn. Wiley, New Jersey

Juvinall R, Marshek K (2011) Fundamentals of machine component design, 5th edn. Wiley, New York

Mott RL, Vavrek EM, Wang J (2018) Machine elements in mechanical design, 6th edn. Pearson Education, New York

Chapter 18
Chain Drives

Nomenclature

A_p	Chain bearing area, mm^2
a	Center distance, mm
b	Lubrication factor
d	Shaft diameter, mm
F	Force, kg
H	Chain life, hours
K	Service factor
k_m	Strand factor
L	Span length of chain, mm
m	M.Y method correction factor
N	Number of sprockets
N_s	Number of strands
n_1	Rotational speed of small sprocket, rpm
n_2	Rotational speed of large sprocket, rpm
P	Transmitted power, kW
P_t	Equivalent power, kW
p_0	Allowable bearing pressure, MPa
p'	Chain pitch, mm
q	Mass of chain per unit length, Kg/m
S	Sag, mm
SF	Safety factor
S_{min}	Minimum allowable sag, mm
u	Speed ratio
v	Linear speed of sprockets, m/s
z_1	Teeth number of small sprocket
z_2	Teeth number of large sprocket
α	Angle of the chain drive with horizon, degrees

M. Yaghoubi and H. Tavakoli, *Mechanical Design of Machine Elements by Graphical Methods*, Materials Forming, Machining and Tribology,
https://doi.org/10.1007/978-3-031-04329-1_18

18.1 Introduction

A chain drive is another common type of flexible drives. It is used for power transmission between shafts that are relatively far from each other. A chain drive includes a driving sprocket (a toothed wheel), a driven sprocket and a chain loop. The chain consists of a series of pin-connected links, which provides the flexibility. During the power transmission, the chain engages the sprockets. Chain drives are typically employed for low to moderate speeds, high torque transmission, and applications in which precise speed ratios must be maintained (Mott et al. 2018; Jiang 2019).

Roller chain is the most popular type of chains, in which the rollers on each pin rotate during their contacts with the teeth of sprockets and provide low friction between the chain and the sprockets, and as a result a high transmission efficiency (Fig. 18.1A). Roller chains are available in single- or multiple-strand types. For higher power transmissions, multiple-strand roller chains are required (Jiang 2019).

Bush chain is another type of chain (Fig. 18.1B). The main difference between roller and bush chains is that the latter does not have rollers. A comparison between the structures of the two chains is provided in Fig. 18.1 (Autodesk Inc. 2021). In this chapter, we discuss the design of both roller and bush chains.

18.2 Design Methods

The design of chain drives usually includes determining size of chain (pitch), the chain type, number of strands, and the teeth number of sprockets.

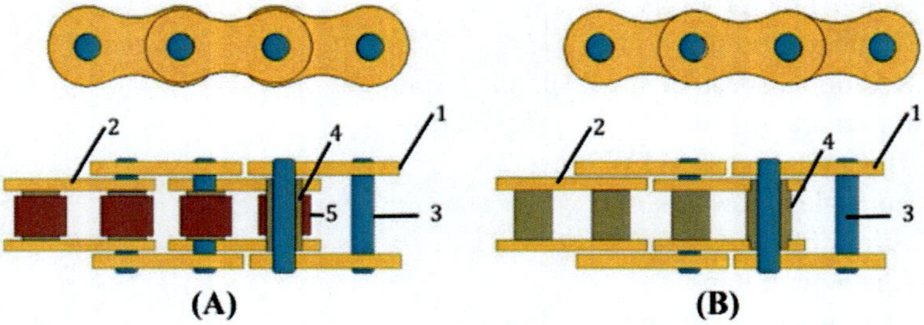

Fig. 18.1 Structure of **A** roller chain and **B** bush chain (1—outer plate, 2—inner plate, 3—bearing pins, 4—bush, and 5—roller) (modified from Autodesk Inc. 2021)

18.2.1 Rough Design

Fig. 18.2 can be employed for rough design of chain drives. The following example illustrates the use of the method.

Example 18.1 A power of $P = 60$ kW at a rotational speed of $n_1 = 600$ rpm is supposed to be transmitted from a small sprocket to a large sprocket (Fig. 18.3). The speed ratio is $u = 4$ and the center distance is $a = 2000$ mm. Determine size of the chain (the chain pitch)? Select a triple-strand chain.

Solution

1 For safety, we multiply the power by 2:

$$P = 2 \times 60 = 120\,\text{kW}$$

2 First, from Fig. 18.2D, we determine the teeth number of small sprocket (z_1). From the horizontal axis, we select the speed ratio of $u = 4$, go up to reach the curve, and then go left to obtain the recommended value for the teeth number:

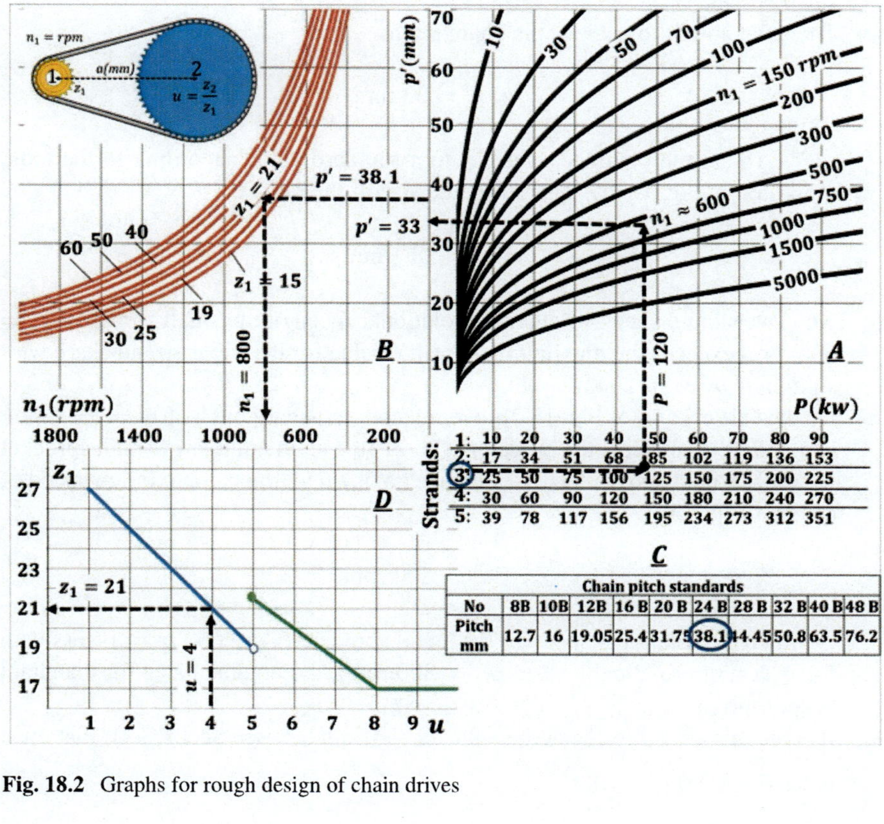

Fig. 18.2 Graphs for rough design of chain drives

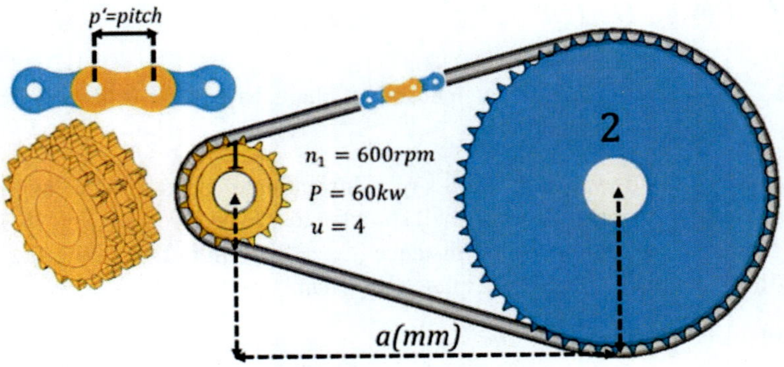

Fig. 18.3 A schematic of the chain drive of Example 18.1

$$z_1 = 21$$

3 Now we find the chain pitch using Fig. 18.2A. As shown, from the horizontal axis, we select the power of $P = 120$ kW for a triple-strand chain, then go up to intersect the approximate curve related to $n_1 = 600$ (the rotational speed of the small sprocket).

4 Then, we go left to achieve the chain pitch:

$$p' = 33 \text{ mm}$$

We round the obtained value up to a standard value according to the table presented in Fig. 18.2C. Therefore, the size of the pitch is:

$$p' = 38.1 \text{ mm}$$

5 Now, we should check whether the rotational speed of the small sprocket ($n_1 =$ 600 rpm) exceeds the maximum allowable value for the triple-strand chain with a pitch of $p' = 38.1$ mm or not.

6 On the vertical axis of Fig. 18.2B, we select the standard pitch of $p' = 38.1$ mm, and go left to intersect the curve of $z_1 = 21$ (the teeth number of small sprocket which was obtained in step 2). Then, we go down to find the maximum speed of the small sprocket:

$$n_1 = 800 \text{ rpm}$$

Since the rotational speed of the small sprocket ($n_1 = 600$ rpm) is less than the maximum rotational speed ($n_1 = 800$ rpm), the obtained size for the chain is appropriate.

7 It is recommended to check the value of $\frac{a}{p'}$ (Eq. 18.1 (see Sect. 18.4)) that must be less than 80 $\left(\frac{a}{p'} \leq 80\right)$:

$$\frac{a}{p'} = \frac{2000}{38.1} = 52.5 < 80$$

Therefore, the designed chain is suitable for this application.

8 A summary of the characteristics of the designed chain:
The chain type: 24B
The number of strands: $N_s = 3$
The chain pitch: $p' = 38.1$ mm.
The teeth number of the small sprocket: $z_I = 21$
The teeth number of the large sprocket (Eq. 18.2 (see Sect. 18.4)):

$$z_2 = uz_1 = 4 \times 21 = 84$$

The rotational speed of the large sprocket (Eq. 18.2 (see Sect. 18.4)):

$$n_2 = \frac{n_1}{u} = \frac{600}{4} = 150 \text{ rpm}$$

18.2.2 Accurate Design

Equation 18.1 (see Sect. 18.4) is usually used for calculating the equivalent power and design of chain drives. However, graphs presented in Figs. 18.4, 18.5 and 18.6 can be applied for easier designing. The following example shows the procedure of employing the graphs.

Example 18.2 A power of $P = 20$ kW at a speed of $n_I = 600$ rpm is to be transmitted by a chain from a small sprocket to a larger one with a speed ratio of $u = 4$. Specify the size of the chain (the chain pitch).

Assumptions: Select a triple-strand chain of type A. Consider the working condition and lubrication state of the chain as "well lubricated, and partially dusty environment" (the lubrication factor: $b = 2$, according to Eq. 18.1 (see Sect. 18.4)). Assume that the service factor is $K = 2$.

Solution

1 First, we multiply the power by the lubrication factor ($b = 2$) and the service factor ($K = 2$):

$$P = 20 \times K \times b \xrightarrow[b=2]{K=2} P = 80\,\text{kW}$$

2 According to Fig. 18.2D, since $u = 4$, we choose the teeth number of small sprocket as:

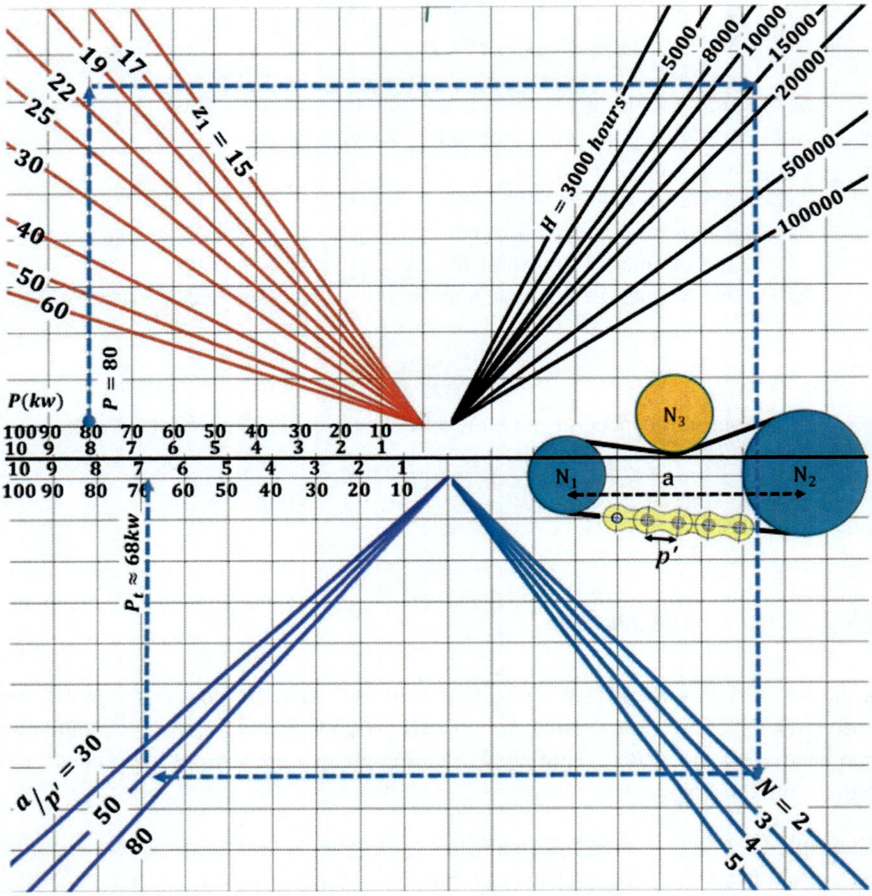

Fig. 18.4 Graphs for determining the equivalent power (P_t) of chain drives

$$z_1 = 21$$

Now, we should calculate the equivalent power (P_t).

3 In Fig. 18.4, from the left horizontal axis, we select the power of $P = 80$ kW and go up to reach the approximate curve related to $z_1 = 21$.

4 Then, we go right to cross the curve of $H = 15{,}000$ h. Arbitrarily, we choose 15,000 h for the chain life.

5 Since the chain drive of this example includes two sprockets (the small one as the driving sprocket and the large one as the driven sprocket), we go down to intersect the curve related to $N = 2$.

6 After that, we go left to reach the curve related to $\frac{a}{p'} = 50$. As the center distance (a) is unknown, we arbitrarily choose $\frac{a}{p'} = 50$.

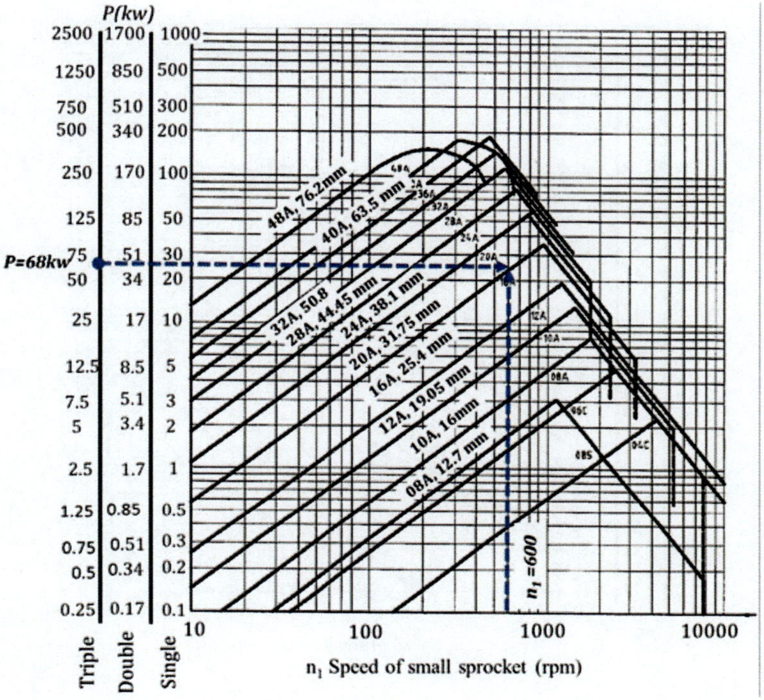

Fig. 18.5 Relationship between the transmitted power and the rotational speed of small sprocket for roller chains (type A) (modified (Decker and Kabus (2014) Maschinenelemente. Carl Hanser Verlag München))

Note: To find the minimum value for the ratio, $\frac{a}{p'}$, you can refer to Example 18.5.

7 Then, we go up to achieve the equivalent power:

$$P_t = 68\,\text{kW}$$

8 Now, in Fig. 18.5, which is related to the A type chains, we select the power of $P = 68$ kW from the part related to triple-strand sprockets on the vertical axis, and go right.

9 At the same time, we choose the rotational speed of $n_1 = 600$ rpm from the horizontal axis, and go up.

The location of the intersection specifies the type of chain. Since the intersection point is on the border between the areas associated with 16A and 20A, we choose the bigger one. Therefore, the chain type is:

$$20\text{ A with a pitch of } p' = 31.75\text{ mm}$$

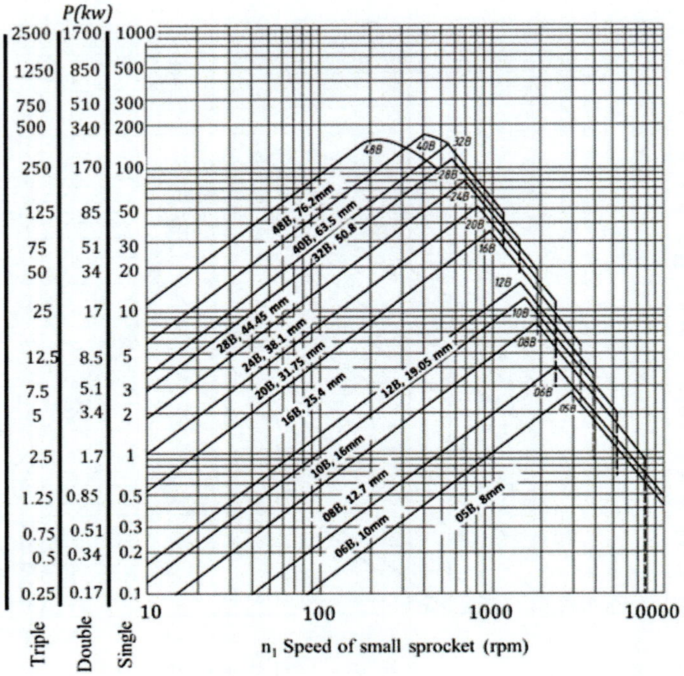

Fig. 18.6 Relationship between the transmitted power and the rotational speed of small sprocket for bush chains (type B) (modified from Wittel et al. 2013)

10 In the end, the specifications of the chain drive are as follows:
The chain type: 20A
The number of strands: $N_s = 3$
The chain pitch: $p' = 31.75$ mm.
The teeth number of the small sprocket: $z_1 = 21$
The teeth number of the large sprocket (Eq. 18.2 (see Sect. 18.4)):

$$z_2 = uz1 = 4 \times 21 = 84$$

The rotational speed of the large sprocket (Eq. 18.2 (see Sect. 18.4)):

$$n_2 = \frac{n_1}{u} = \frac{600}{4} = 150\,\text{rpm}$$

The center distance:

$$\frac{a}{p'} = 50 \rightarrow a = p' \times 50 = 1587.5\,\text{mm}$$

The maximum center distance:

$$\frac{a}{p'} \leq 80 \rightarrow a \leq p' \times 80 \rightarrow \mathrm{a} \leq 2540\,\mathrm{mm}$$

18.2.3 Sag Adjustment for Roller Chains

Proper sag in chain drives is very important. To determine the appropriate amount of sag, you can refer to one of the methods described in the following subsections.

18.2.3.1 Method of Renold plc

Based on this method, for vertical chain drives (Fig. 18.7A), the total movement of chain ($2S$) should be approximately equal to half the chain pitch $\left(p'/2\right)$. While, for horizontal chain drives (Fig. 18.7B), the total movement of chain ($2S$) should be equal to 4% of the center distance (a/25) in case of smooth drives, and equal to 2% of the center distance (a/50) in case of shock drives (Eq. 18.3 (see Sect. 18.4)) (Renold plc 2020).

18.2.3.2 Method of Daido Kogyo Co.

According to this method, in general, we can consider sag as about 2% of the span length ($S = 0.02L$) (Fig. 18.8). However, under each of the conditions below, it is

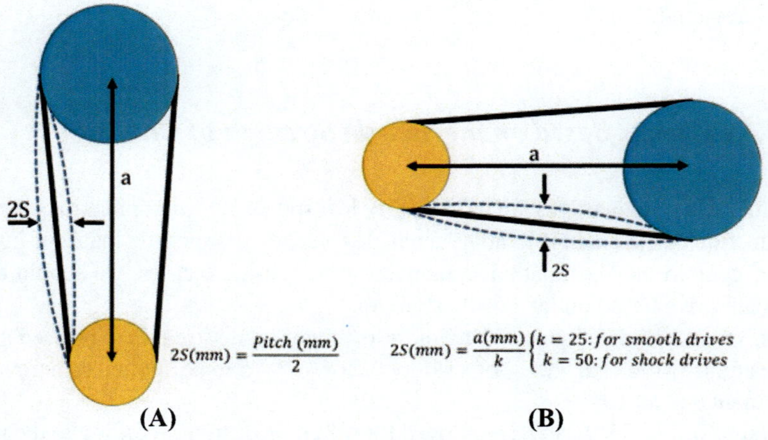

Fig. 18.7 An illustration of chain sag according to Renold plc for **A** vertical chain drive, and **B** horizontal chain drive

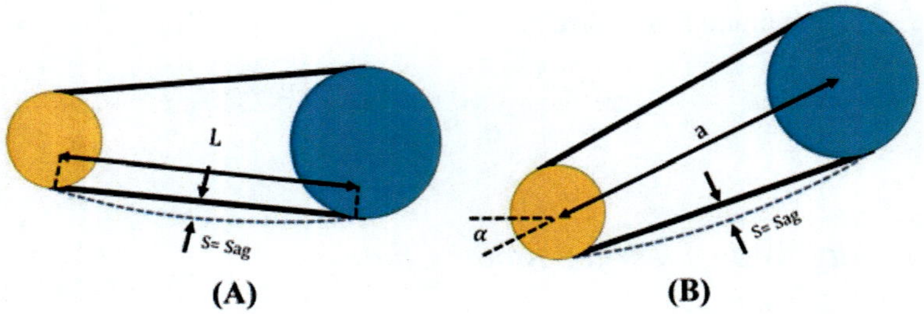

Fig. 18.8 an illustration of chain sag, **A** according to Daido Kogyo Co., and **B** according to Zhukov method

recommended to choose a value equal or less than about 1% of the span length ($S \leq 0.01L$) (Daido Kogyo Co 2007):

- In the case that the chain is installed vertically or semi vertically.
- In the case that the chain is installed horizontally or semi horizontally with the top slackened.
- In the case that the center distance exceeds 50 times the chain pitch.
- In the case that vibration or shock occurs.
- In the case that the chain is frequently started and stopped.
- In the case that the chain is suddenly reversed.
- In the case that the speed ratio is 7 or more ($u \geq 7$) (however, keeping the speed ratio at 7 or less is safer and preferable).

18.2.3.3 Method of Zhukov

In this method, the chain sag can be calculated using Eq. 18.4 (see Sect. 18.4) (Zhukov and Gurevich 2014).

18.2.4 Design Based on the Tensile Strength of Chains

In some cases, chain drives are used in very low speeds, or for applications in which the function of the chain is to apply a tensile force, or to support a load; for instance in conveyors or devices that have many sprockets. In such cases, we should design the chain drive based on the tensile strength.

It is recommended that only 10% of the average tensile strength to be used in such applications (Mott et al. 2018) or a safety factor of 7–15 to be applied (Anurev 2006; Bakumenko et al. 1997).

Graphs of Fig. 18.9 can be employed for the chain design according to the tensile strength. For more details refer to the following example.

Example 18.3 In Fig. 18.10, a conveyor is shown that moves an object continuously

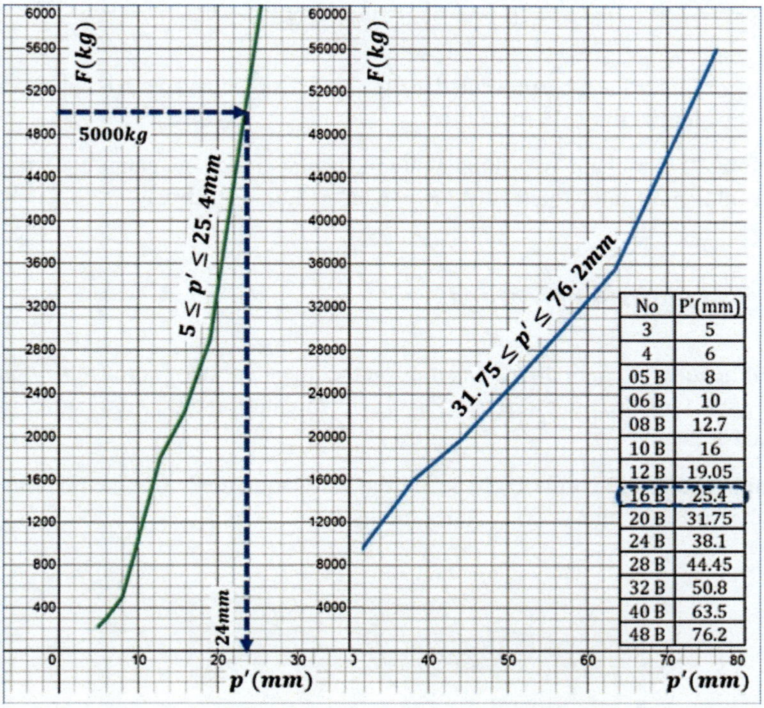

Fig. 18.9 Tensile strength of chains

Fig. 18.10 An illustration for the conveyor of Example 18.3

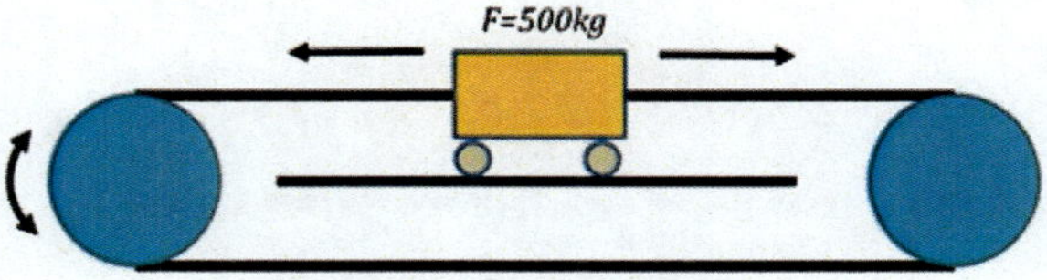

left and right. Assume that the resultant force of acceleration, friction force, etc. is equal to $F = 500$ kg and the chain is single-strand. Specify the type of chain.

Solution:

1. First, we multiply the force (F) by a safety factor (SF) of 10:

$$F = SF \times 500 \xrightarrow{SF=10} F = 5000\,\text{kg}$$

2. Now, in Fig. 18.9, we find the force of $F = 5000$ kg on the vertical axis and go right to intersect the curve.
3. Then, we go down to obtain the chain pitch:

$$p' = 24 \text{ mm}$$

4. According to the table presented in Fig. 18.9, the obtained pitch is approximately equivalent to the 16B chain type.

   Note: If the chain was double-strand, triple-strand, or quadruple-strand, we would apply safety factors of $SF = 6, 4,$ or 3.4, respectively.

18.2.5 Lubrication Method

The lubrication of chain drives depends on the linear speed of sprockets (v) and the chain pitch (p'). You can refer to Fig. 18.11 to determine the lubrication method (Autodesk Inc. 2021).

Example 18.4 In Example 18.2 specify the lubrication method. $n_1 = 600$ rpm, $z_1 = 21$ and $p' = 31.75$ mm.

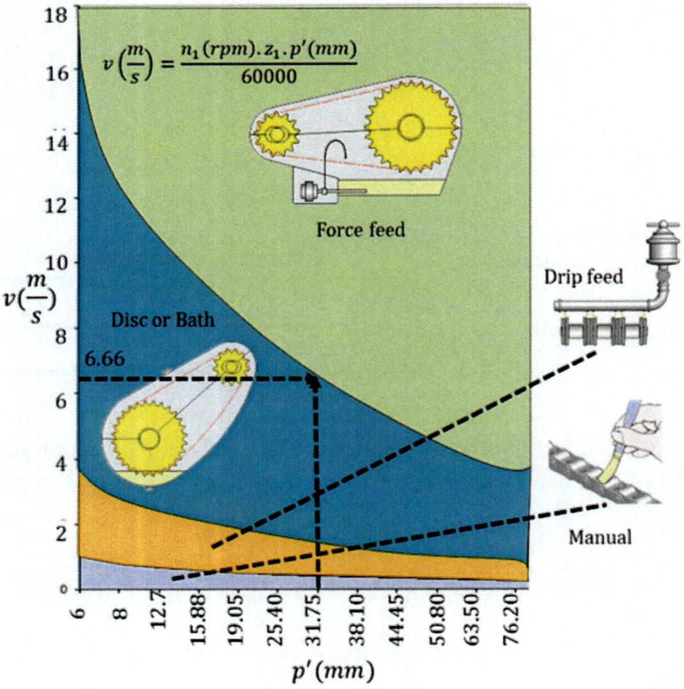

Fig. 18.11 A diagram for determining the lubrication method of chain drives (modified from Autodesk Inc. 2021)

Solution

1. First, from the equation shown in Fig. 18.11, we calculate prophetical speed of the chain:

$$v = \frac{n_1 z_1 p'}{60000} \xrightarrow{n_1=600, z_1=21, p'=31.75} v = \frac{600 \times 21 \times 31.75}{60000} = 6.66\,\mathrm{m/s}$$

2. In Fig. 18.11, from the horizontal axis, we select $p' = 31.75$ mm and go up; simultaneously from the vertical axis, we select $v = 6.66$ m/s and go right. The intersection point is on the border of "Forced feed" and "bath or disc" lubrication. Thus it is better to choose "Forced feed" lubrication method.

18.2.6 M.Y Method

This method is applicable when the shaft diameter is known and there is no time for designing the chain drive. In this case, the chain pitch can be estimated based on the shaft diameter (d) as shown in Table 18.1 and also in Sect. 18.4 (Eq. 18.5). In addition, the maximum allowable speed of small sprocket can be calculated according to the estimated pitch (Table 18.1 and Eq. 18.6 (see Sect. 18.4)). Note that in this method, the teeth number of small sprocket must be $z_1 \geq 17$.

Example 18.5 In Example 18.1, the power of $P = 60$ kW at the rotational speed of $n_1 = 600$ rpm was supposed to be transmitted from the small sprocket to a large sprocket. The speed ratio was $u = 4$. Find the chain pitch using M.Y method. Assume that the chain is triple-strand the shaft arrangement is cantilever with a diameter of $d = 90$ mm.

Table 18.1 Relationships of M.Y method

Shaft arrangement	Symmetrical or asymmetrical 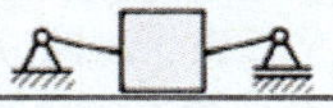	Cantilever 
Chain pitch (p')	$\begin{cases} 0.8md & (\text{for } n_1 > 630\,\text{rpm}) \\ 0.6md & (\text{for } n_1 \leq 630\,\text{rpm}) \end{cases}$	$\begin{cases} 0.6md & (\text{for } n_1 > 630\text{ rpm}) \\ 0.5md & (\text{for } n_1 \leq 630\text{ rpm}) \end{cases}$
Maximum rotational speed of small sprocket ($z_1 = 17$)	$n_1 \leq \frac{28000}{p'}$	
M.Y method correction factor (m)	$\begin{cases} 1 & (\text{for single-strand chain}) \\ 0.85 & (\text{for double-strand chain}) \\ 0.75 & (\text{for triple-strand chain}) \\ 0.7 & (\text{for quadruple-strand chain}) \end{cases}$	

Solution

1 For safety, we multiply the power by 2:

$$P = 2 \times 60 = 120\,\text{kW}$$

2 Since $n_1 < 630$ rpm, the chain is triple-strand, and the shaft arrangement is cantilever, according to Table 18.1, the pitch size is:

$$p' = 0.5\text{md} \quad \xrightarrow[]{\substack{m = 0.75 \\ d = 90}} \quad \text{p}' = 0.5 \times 0.75 \times 90 \approx 33.75\,\text{mm}$$

We round it up to a higher standard value, which is equivalent to a 24A type chain with $p' = 38.1$ mm.

As you can see, the result is the same as the one achieved in Example 18.1.

3 The maximum rotational speed of small sprocket with $z_1 = 17$ is:

$$n_1 \le \frac{28000}{p'} = \frac{28000}{38.1} = 735\,\text{rpm}$$

Note: If the teeth number increases, the allowable rotational speed increases too.

18.2.7 Estimating the Minimum Center Distance

The minimum center distance can be estimated according to the speed ratio (u) and the teeth number of small pulley (z_1) using Fig. 18.12.

Example 18.6 Consider two power transmission systems that are using chain drives. The system I has a small sprocket with a teeth number of $z_1 = 17$ and a speed ratio of $u = 2$. The system II has a small sprocket with a teeth number of $z_1 = 29$ and a speed ratio of $u = 8$. Determine the minimum center distance for each of the systems.

Solution

1 From the horizontal axis of Fig. 18.12, we select $u = 2$ and 8, and go vertically up.

2 At the same time, from the vertical axis, we select $z_1 = 17$ and 29, and go horizontally right so that the vertical and horizontal lines of each system intersect each other as shown in Fig. 18.12.

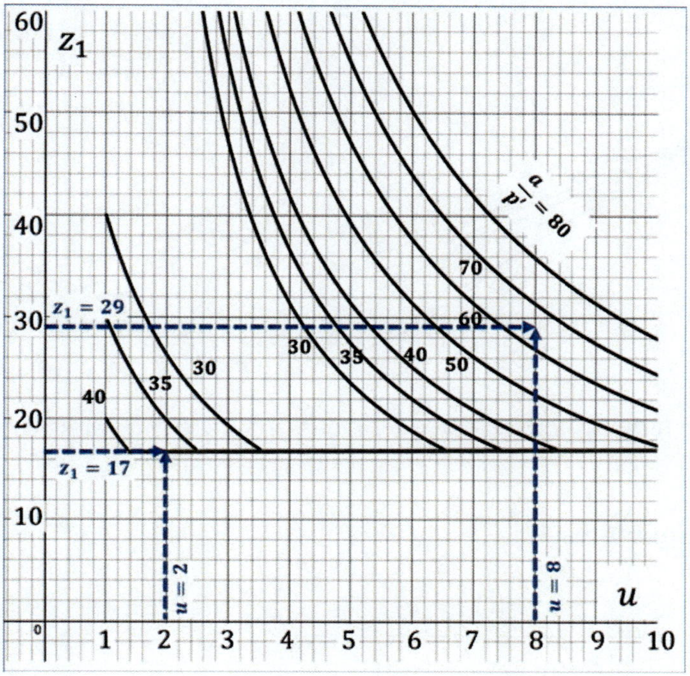

Fig. 18.12 A graph for estimating the minimum value of $\frac{a}{p'}$

3 As we can see, for the system I, the intersection point is between the curves of $\frac{a}{p'} = 40$ and $\frac{a}{p'} = 35$, and for the system II, the intersection point is between the curves of $\frac{a}{p'} = 70$ and $\frac{a}{p'} = 60$.

We should take the maximum values. Thus, as a result, the minimum center distance for the two systems is:

For the system I: $\frac{a}{p'} \geq 40 \rightarrow a \geq 40\,p'$.

For the system II: $\frac{a}{p'} \geq 70 \rightarrow a \geq 70\,p'$.

Note: If you determine the value of z_1 from Fig. 18.2D, there is no need to use Fig. 18.12. You just should note that the value of the center distance should be more than $40p'$ $\left(a \geq 40p'\right)$.

18.3 Supplementary Graphs and Equations

If you intend to check the pressure on the chain (the bearing pressure, p_O), and calculate the chain bearing area (A_p), you can use graphs of Fig. 18.13 and Eq. 18.7 (see Sect. 18.4), respectively.

In addition, equations used for Figs. 18.2 and 18.9 can be seen in Eqs. 18.8 and 18.9, respectively (see Sect. 18.4).

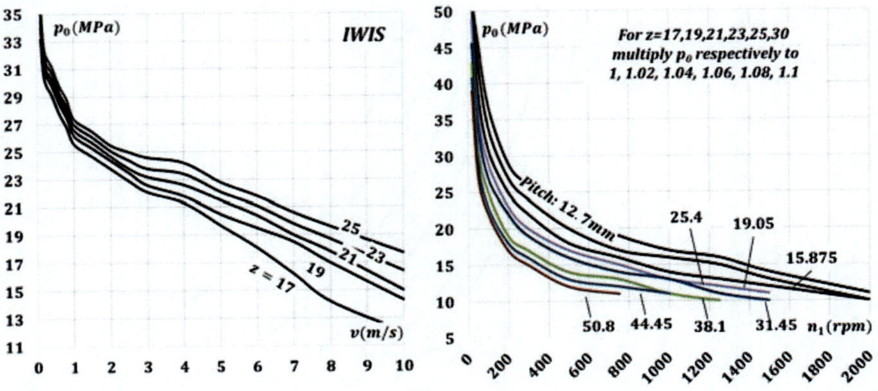

Fig. 18.13 The permissible bearing pressure (p_O) of chains as a function of their linear speed (v) and the rotational speed of small sprockets (n_I)

18.4 Equations

$$P_t = \frac{P\left(\frac{19}{z_1}\right)^{1.08} b}{\left(\frac{15000}{H}\right)^{\frac{1}{3}} (0.9)^{N-2} \times 0.45\left(\frac{a}{p'}\right)^{0.215}}$$

$$\text{Recommended value for } \frac{a}{p'}: 30 - 50, \text{ and maximum value} : 80$$

$$b = \begin{cases} 1 & \text{(for well lubricated, and clean environment)} \\ 2 & \text{(for well lubricated, and partially dusty environment)} \\ 5 & \text{(for insufficiently lubricated, and dusty environment)} \\ 6 & \text{(for non-lubricated, and dusty environment)} \end{cases} \tag{18.1}$$

$$u = \frac{z_2}{z_1} = \frac{n_1}{n_2} \tag{18.2}$$

The recommended sag for chain drives—method of Renold plc (2020):

$$\begin{aligned} &\text{For vertical chain drives}: \\ &2S = \frac{p'}{2} \\ &\text{For horizontal chain drives}: \\ &2S = \begin{cases} \frac{a}{25} & \text{(for smooth drives)} \\ \frac{a}{50} & \text{(for shock drives)} \end{cases} \end{aligned} \tag{18.3}$$

The recommended sag for chain drives—method of Zhukov and Gurevich (2014):

$$S_{min} \leq S \leq 3S_{min}$$

$$S_{min} = \frac{11.4\sqrt[2]{a^3}}{k}\cos\alpha$$
$$k = \begin{cases} 1 & \left(\text{if } v < 10\frac{m}{s}\right) \\ 0.7 & \left(\text{if } v \geq 10\frac{m}{s}\right) \end{cases}$$
$$v = \frac{z_1 n_1 p'}{6000} \tag{18.4}$$

The equations of M.Y method for estimating the chain pitch:

$$For\ symmetrical\ or\ asymmetrical\ shaft\ arrangement:$$
$$p' = \begin{cases} 0.8md & (\text{for } n_1 > 630\,\text{rpm}) \\ 0.6md & (\text{for } n_1 \leq 630\,\text{rpm}) \end{cases}$$
$$For\ cantilever\ shaft\ arrangement:$$
$$p' = \begin{cases} 0.6md & (\text{for } n_1 > 630\,\text{rpm}) \\ 0.5md & (\text{for } n_1 \leq 630\,\text{rpm}) \end{cases}$$
$$m = \begin{cases} 1 & (\text{for single-strand chain}) \\ 0.85 & (\text{for double-strand chain}) \\ 0.75 & (\text{for triple-strand hain}) \\ 0.7 & (\text{for quadruple-strand chain}) \end{cases}$$
$$z_1 \geq 17 \tag{18.5}$$

The equation of M.Y method for estimating the maximum rotational speed of small sprocket:

$$n_1 \leq \frac{28000}{p'}$$
$$z_1 = 17 \tag{18.6}$$

Chain bearing area (A_p):

$$\text{Type A}:$$
$$A_p = 0.27\,p'^2$$
$$\text{Type B}:$$
$$A_p = 0.35\,p'^2 \tag{18.7}$$

For double- and triple-strand, multiply A_p by 2 and 3, respectively.
The equations of Fig. 18.2:

$$n_{max} = 14000\frac{\sqrt[4]{z_1}}{p'}$$

$$p' \geq 30.5\sqrt[3]{\frac{P}{k_m n}} \times \max\left(0.1n_1^{\frac{2}{3}},\, 10(0.1n_1)^{\frac{1}{9}}\right)$$

$$z_1 = \begin{cases} 29 - 2u & (\text{for } u \leq 5) \\ 29 - 1.5u & (\text{for } u > 5) \end{cases}$$

$$k_m = \begin{cases} 1 & (\text{for single-strand chain}) \\ 1.7 & (\text{for double-strand chain}) \\ 2.5 & (\text{for triple-strand chain}) \\ 3 & (\text{for quadruple-strand chain}) \end{cases} \tag{18.8}$$

The equations of Fig. 18.9:

$$\begin{aligned} &\text{Type A}: \\ &F_{max} = 8.78\, p'^{\,2} \\ &q = 0.0039\, p'^{\,2} \\ &\text{Type B}: \\ &F_{max} = 9.5\, p'^{\,2} \\ &q = 0.004\, p'^{\,2} \end{aligned} \tag{18.9}$$

For double- and triple-strand, multiply the values by 2 and 3, respectively.

References

Anurev VI (2006) Handbook of the designer-mechanical engineer. Mechanical Engineering, Moscow (in Russian)

Autodesk Inc. (2021) Engineer's handbook. Retrieved from Autodesk Inventor support and learning center: https://knowledge.autodesk.com/

Bakumenko VI, Bondarsnko VA, Koserukov SN (1997) A short guide to the designer of non-standard equipment. Mechanical Engineering, Moscow (in Russian)

Daido Kogyo Co. (2007) General catalog—power transmission & conveyor chain. Retrieved from https://www.did-daido.co.jp/documents/en/catalog/didcatalog.pdf

Decker K-H, Kabus K (2014) Decker Maschinenelemente Funktion, Gestaltung und Berechnung. Carl Hanser Verlag, München (in German)

Jiang W (2019) Analysis and design of machine elements, 1st edn. Wiley, New Jersey

Mott RL, Vavrek EM, Wang J (2018) Machine elements in mechanical design, 6th edn. Pearson Education, New York

Renold plc (2020). Renold Roller Chain Catalogue. Retrieved from Renold plc: https://www.renold.com/downloads/chain-brochures-and-downloads/

Wittel H, Muhs D, Jannasch D, Voßiek J (2013) Roloff/Matek Maschinenelemente Normung, Berechnung, Gestaltung. Springer Vieweg, Wiesbaden (in German)

Zhukov KP, Gurevich YE (2014) Design of parts and assemblies of machines: a textbook for universities. Mechanical Engineering, Moscow (in Russian)

Chapter 19
Spur and Helical Gear Drives

Nomenclature

a	Center distance, mm
b	Teeth face width (gear width), mm
D_G	Pitch circle diameter of gear, mm
D_p	Pitch circle diameter of pinion, mm
d	Shaft diameter, mm
H_G	Gear hardness
H_p	Pinion hardness
HB	Brinell hardness scale
HRB	Rockwell hardness scale B
HRC	Rockwell hardness scale C
HV	Vickers hardness scale
k, $k_0 - k_5$, k_A, $k_{H\beta}$	Factors for the equations
m	Module, transverse module of helical gears, mm
m_n	Normal module, mm
n_G	Rotational speed of gear, rpm
n_p	Rotational speed of pinion, rpm
T	Torque, N m
T_1	Torque on the shaft of pinion, N m
T_E	Maximum torque of engine, N m
u	Speed ratio, transmission ratio
$u_1, u_2,$ $and\, u_3$	Transmission ratios of gear stages in a multi-stage gearbox
u_t	Overall transmission ratio of a multi-stage gearbox

M. Yaghoubi and H. Tavakoli, *Mechanical Design of Machine Elements by Graphical Methods*, Materials Forming, Machining and Tribology,
https://doi.org/10.1007/978-3-031-04329-1_19

z_G	Number of gear teeth
z_p	Number of pinion teeth
β	Helix angle, degrees
σ_F	Allowable bending stress, MPa
σ_H	Allowable contact stress, MPa
φ	Ratio of pinion width to its pitch circle diameter

19.1 Introduction

Gears are toothed wheels that are widely used for power and motion transmission between rotating shafts. Gear drive, also known as gearing, refers to an assembly of two or more gears. However, a typical gear drive is composed of a small gear, which is termed as pinion and a larger one, which is called a gear. Compared with other means of mechanical power transmission, such as belt drives and chain drives, gear drives possess advantages including high efficiency, reliability and durability, compact design, uniform transmission ratio, precise timing, and transmission of a wide range of speed and speed ratios. On the other hand, manufacturing costs of gears are usually higher than chains and belts. In addition, gear drives are not suitable for the power transmission between widely spaced shafts (Collins et al. 2009; Juvinall and Marshek 2011; Jiang 2019).

Based on the relative position of the shafts, gear drives can be fell into three principal types or categories, namely parallel-shaft gear drives (in which the shaft axes are parallel), intersecting-shaft gear drives (in which the shaft axes intersect) and crossed gear drives (in which the shaft axes are neither parallel nor intersecting) (Collins et al. 2009; Jiang 2019).

Parallel-shaft gear drives are the most common type of drives, because they are simples. Different form of gears are used in Parallel-shaft gearings, however, spur gears and helical gears are more popular. Spur gears have straight teeth that are parallel to the axis of the shaft. Among all types of gears, spur gears are the simplest. While, helical gears have teeth, which are inclined to the axis. There are also gears with double helical teeth that are called herringbone gears. The applications of both spur and helical gears can be the same, although because of the more gradual engagement of the teeth in helical gears, they are less noisy (Jiang 2019).

Intersecting-shaft gearings (bevel gear drives) and crossed gearings (wormgear drives) will be discussed in Chaps. 20 and 21, respectively.

This chapter deals with designing spur and helical gear drives.

19.2 Design Methods

Gearings design requires using complex diagrams and many trials and errors. However, in this chapter, we propose methods that give very good approximations and easy calculations for the design. Since, in a gear drive, the small gear (pinion) is more vulnerable to failure, it is necessary to base the design on it.

To apply our method for designing pinions, the following steps should be taken:

1 First, from Table 19.1 (or Supplementary Table 19.1) and according to the material and hardness of the pinion, calculate the allowable contact stress (σ_H).
2 Then, determine the ratio of the pinion width to its pitch circle diameter (φ) which depends on the arrangement of gears on the shaft, from Table 19.2.
3 Then, find the pitch diameter of pinion (D_p) from Fig. 19.1 (or Eq. 19.1 (see Sect. 19.4)).
4 Using the φ ratio and Eq. 19.2 (see Sect. 19.4), the pinion width (b) (in fact the face width of teeth) can be calculated: $b = D_p \times \varphi$.
5 Then, determine the number of pinion teeth (z_p) using Figs. 19.2, 19.3 and 19.4.
6 Then, calculate the module (m) using Eq. 19.3 (see Sect. 19.4) $\left(m = \frac{D_p}{z_p}\right)$ and round it to a standard module from Table 19.3.
7 In the end, specify the quality of the pinion teeth, which depends on the pitch circle diameter and the velocity of the pinion, using Fig. 19.5. The higher the velocity, the higher the quality should be.

The above mentioned steps are for spur gearings, however, in case of helical gearings, the helix angle should be estimated too. Section 2.4 introduces methods for this purpose.

In addition, for determining the center distance, refer to Sect. 2.5.

Table 19.1 The allowable contact stress (σ_H) for material used in spur and helical gears

Heat treatment	Hardness	Material type	$\sigma_H (MPa)$
Normalized or tempered	<350 HRB	Carbon and alloy steel	2 × HB + 70
Induction hardening	38 – 56 HRC		17×HRC+200
Cemented or carburized	>56 HRC	Alloy steel	23 × HRC
Nitride	550 – 750 HV		1050 – 1150
Without thermal processing	–	Cast iron	400 – 500

Table 19.2 The ratio of pinion width to its pitch circle diameter (φ) in spur and helical gears according to different references

Reference	Wittel et al. (2013)				Zhukov and Gurevich (2014)		Böge (2011)
Shaft arrangement	Normalized Hardness < 180 HB	Tempered Hardness > 200 HB	Induction hardening cemented or carburized	Nitride	Hardness ≤ 350 HB	Hardness > 350 HB	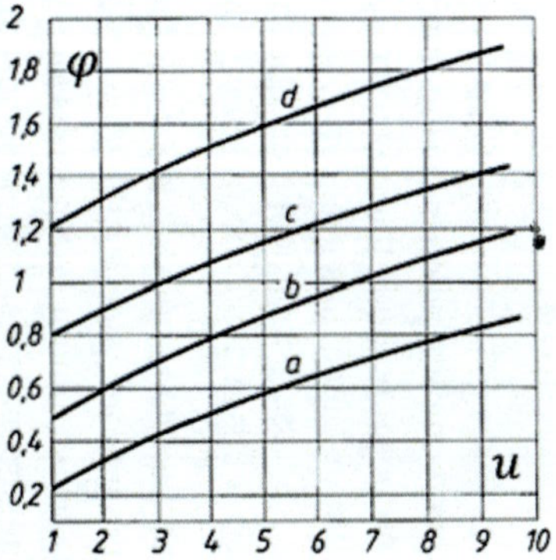
Symmetrical 	$\varphi < 1.6$	$\varphi < 1.4$	$\varphi < 1.1$	$\varphi < 0.8$	$\varphi = 0.6 - 1.2$	$\varphi = 0.4 - 0.8$	
Asymmetrical 	$\varphi < 1.3$	$\varphi < 1.1$	$\varphi < 0.9$	$\varphi < 0.6$	$\varphi = 0.5 - 1.0$	$\varphi = 0.3 - 0.6$	
Cantilever 	$\varphi < 0.8$	$\varphi < 0.7$	$\varphi < 0.6$	$\varphi < 0.4$	$\varphi = 0.3 - 0.4$	$\varphi = 0.25–0.3$	*a*: low speed gearbox, cantilever, car gearbox *b*: medium speed gearbox, general design *c*: high speed and quality *d*: high speed, highest quality and rigidity

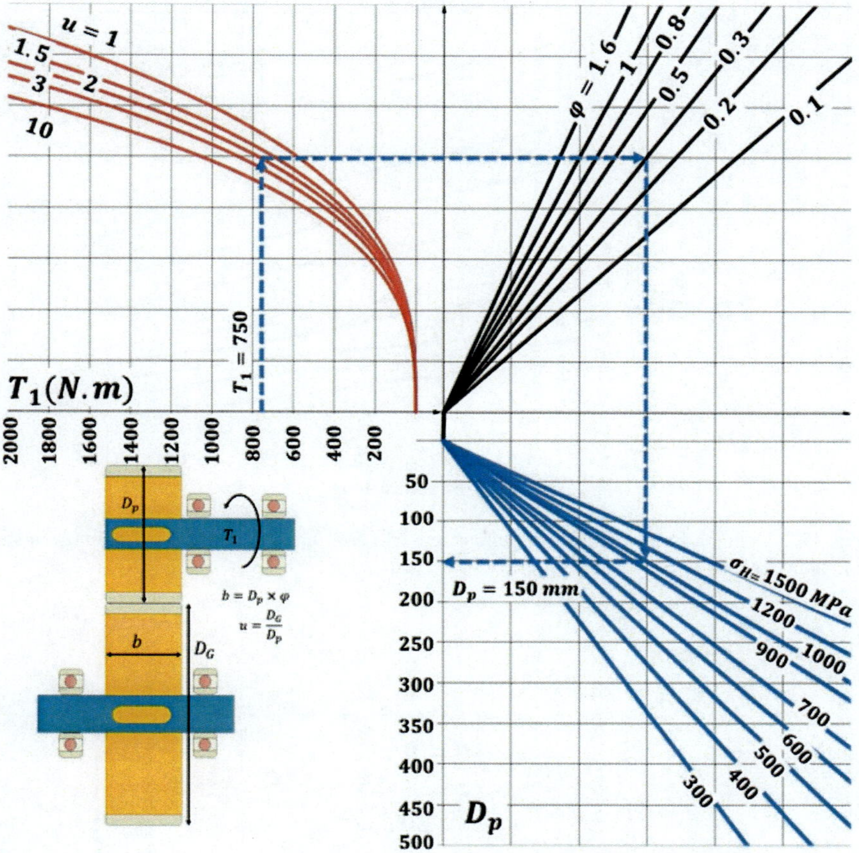

Fig. 19.1 Graphs for determining the pitch circle diameter of pinion (D_p) in spur and helical gearings

19.2.1 *Estimating the Width and Pitch Circle Diameter of Pinion*

As mentioned above, using Tables 19.1, 19.2, Fig. 19.1 and Eq. 19.2 (see Sect. 19.4), the width and pitch circle diameter of pinion can be estimated. The following example shows the process of applying the method.

Example 19.1

Figure 19.6 shows a gear drive together with the values of the torque and rotational speed of the pinion and gear, and also the speed ratio. The pinion teeth have been hardened by the induction hardening method, and the hardness is 50 HRC. Find the size of pinion pitch diameter (D_p) and its width (b).

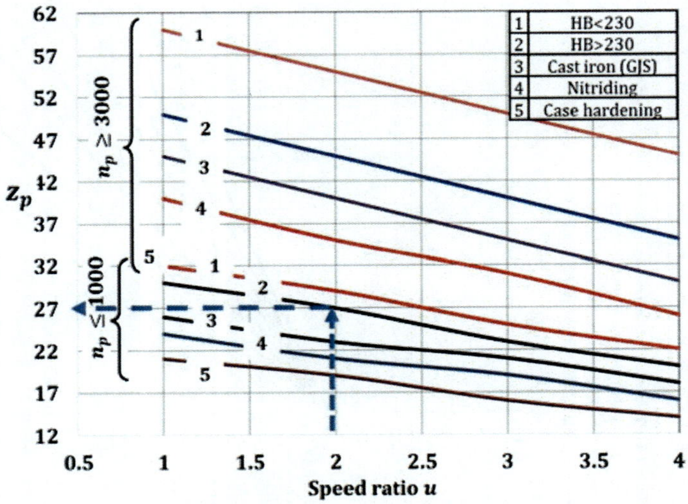

Fig. 19.2 A graph for the number of pinion teeth (z_p) as a function of the speed ratio (u) in spur and helical gears

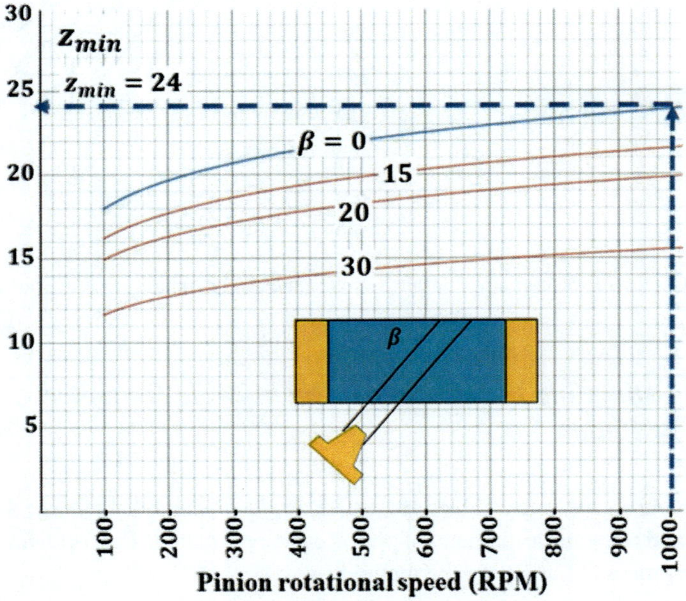

Fig. 19.3 A graph for the minimum number of pinion teeth (z_{min}) as a function of the rotational speed of pinion (n_p) in spur and helical gears

	....											normalized
	(GG-25)	(GTS-35)	(GTS-65)	(GGG-40)	(GGG-60)	(GGG-80)	GS-52	GS-60	E295 (St 50)	E335 (St 60)	E360 (St 70)	C 45
M	0.0008	0.0032	0.0019	0.0027	0.0019	0.0014	0.0027	0.0022	0.0023	0.0019	0.0019	0.0015
	220 HB	150 HB	220 HB	180 HB	250 HB	320 HB	160 HB	180 HB	160 HB	190 HB	210 HB	190 HV 10

....	hardened and tempered			case hardened			gas nitride			nitrocarburized			carbonitrided
	34CrMo4	42CrMo4	34CrNiMo6	16MnCr5	15CrNi6	17CrNiMo6	42CrMo4	16MnCr5	31CrMoV9	C 45	16MnCr5	42CrMo4	34Cr4
M	0.0019	0.0016	0.0015	0.0004	0.0004	0.0004	0.0007	0.0007	0.0006	0.0012	0.0011	0.0010	0.0005
	27HRC	31HRC	32HRC	59HRC	730 HV 10	740 HV 10	52HRC	52HRC	58HRC	43HRC	53HRC	55HRC	56HRC

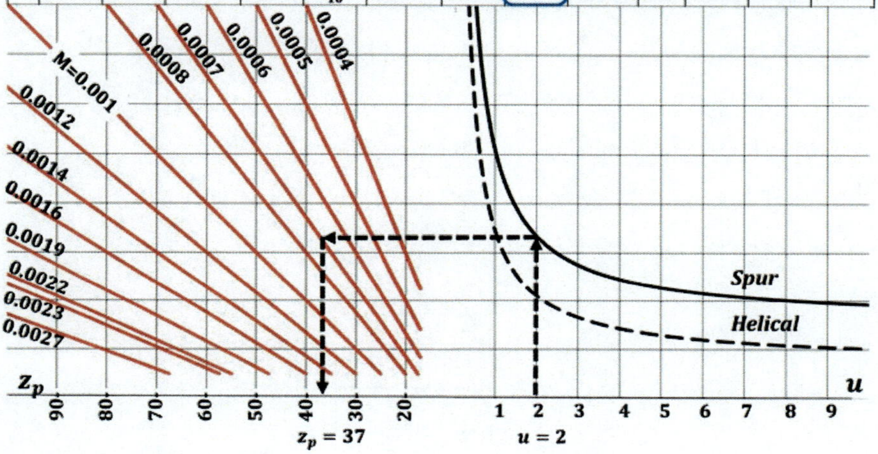

Fig. 19.4 Graphs for determining the number of pinion teeth (z_p) according to the pinion material in spur and helical gears

Solution

1 First, we should determine the value of allowable contact stress (σ_H). Since the pinion teeth have a hardness of 50 HRC and the method of hardening is the induction hardening, according to Table 19.1, the value of σ_H is:

$$\sigma_H = 17 \times \text{HRC} + 200 \rightarrow \sigma_H = 17 \times 50 + 200 \approx 1000\,\text{MPa}$$

2 Now, we select the value of φ ratio from Table 19.2. The arrangement of pinion on its shaft is cantilever (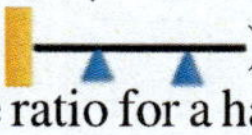). From Table 19.2 and according to Zhukov and Gurevich (2014), the ratio for a hardness of 50 HRC (≈490 HB (Jiang 2019)) is $\varphi = 0.25 - 0.3$. We choose $\varphi = 0.3$.

3 The torque on the pinion is $T_1 = 500$ N m. For safety, we multiply it by 1.5:

$$T_1 = 1.5 \times 500 = 750\,\text{N.m}$$

4 From the left horizontal axis of Fig. 19.1, we select $T_1 = 750$ N m and go up to intersect the curve of the speed ratio $u = 2$.

5 Then, we go right to reach the curve of $\varphi = 0.3$.

6 Then, we go down to cross the curve related to $\sigma_H = 1000$ MPa.

Table 19.3 Standard modules of spur and helical gears (all values are in mm)

Series 1	0.05	0.06	0.08	0.1	0.12	0.16	0.2	0.25	0.3	0.4	0.5	0.6	0.7	0.8	0.9	1	1.25
	1.5	2	2.5	3	4	5	6	8	10	12	16	20	25	32	40	50	60
Series 2	0.055	0.07	0.09	0.11	0.14	0.18	0.22	0.28	0.35	0.45	0.55	0.65	0.75	0.85	0.95	1.125	1.375
	1.175	2.25	2.75	3.5	4.5	5.5	7	9	11	14	18	22	28	36	45	55	70

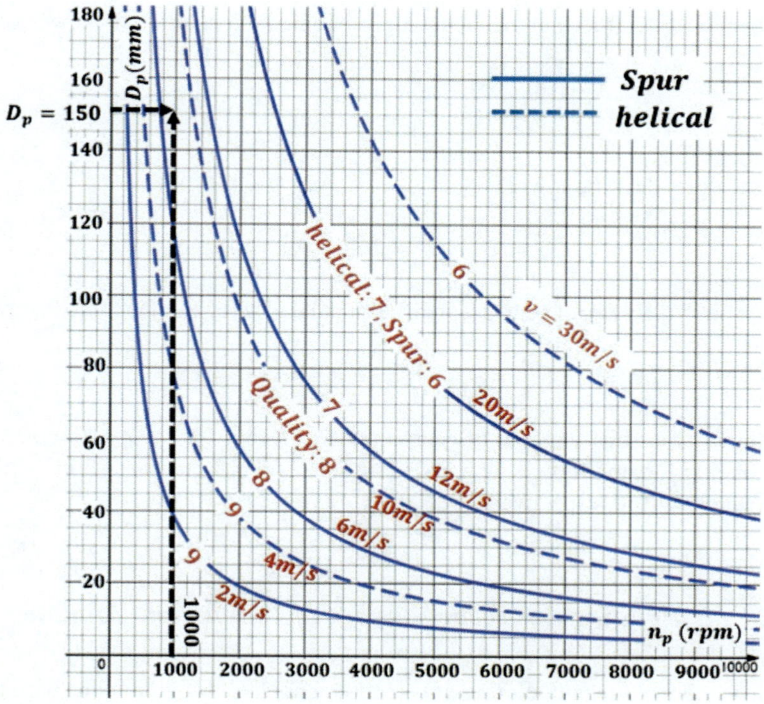

Fig. 19.5 A graph for specifying the gear quality of spur and helical gearings in GOST standard

Fig. 19.6 An illustration for the gear drive of Example 19.1

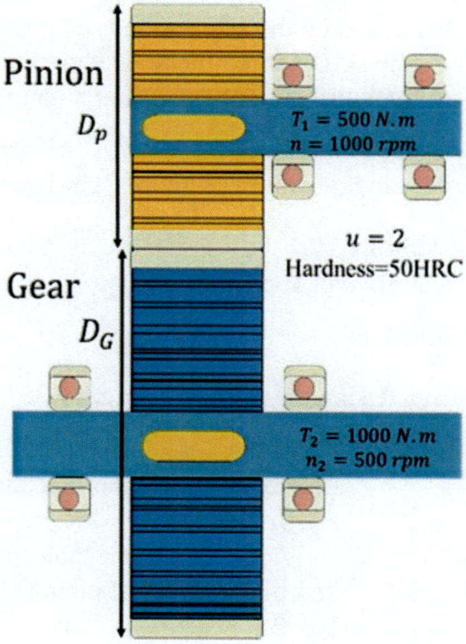

7 Then, we go left to find the value of pinion pitch diameter on the lower vertical axis:

$$D_p = 150\,\text{mm}$$

8 Therefore, the pinion width is:

$$b = D_p \times \varphi = 150 \times 0.3 = 45\,\text{mm}$$

9 The pitch diameter of gear (D_G) can be calculated using Eq. 19.3 (see Sect. 19.4):

$$D_G = u \times D_p = 2 \times 150 = 300\,\text{mm}$$

Note: in cases that material of gear is plastic, in Fig. 19.1, use σH = 300 MPa. After finding Dp, multiply it by 2.7, 2, 1.4, and 2 for the material GFPOM, POM, PA66, and GFPA66 respectively.

19.2.2 Estimating the Teeth Number of Pinion, and Module

For an economical manufacturing, it is recommended to design pinions with higher number of teeth and smaller module (Decker and Kabus 2014). The module size should be $m \geq 1$ mm for spur gears and $m \geq 1.5$ mm for helical gears. The designer is responsible for choosing the number of pinion teeth. However, it should be greater than z_{min}, and the suggestion is $z_p \geq 26$ (Feshenko 2016).

Any of the three methods shown in Figs. 19.2, 19.3 and 19.4 can be used for estimating the number of pinion teeth. After the estimation, the module of pinion can be calculated using Eq. 19.3 (see Sect. 19.4) $\left(m = \frac{D_p}{z_p}\right)$ and be selected based on the standard values of Table 19.3. For more details, refer to the following example.

Example 19.2 Determine the number of pinion teeth (z_p) and module (m) for the gear drive of Example 19.2.

Solution

(a) Using the method of Fig. 19.2:

1 From the horizontal axis of Fig. 19.2, we choose the speed ratio of $u = 2$, and go up to intersect the curve No. 2 of the rotational speed range $n_p \leq 1000\,\text{rpm}$.

(According to the table presented in the upper right corner of Fig. 19.2, and the hardness of pinion (50 HRC ≈ 490 HB), we should use the curve No. 2.)

2 Then, we go left to obtain the recommended number of pinion teeth:

$$z_p = 27$$

Note: If the rotational speed of pinion is between 1000 and 3000 rpm (for example, $n_p = 2000$ rpm), you should find a value for z_p using the curve No. 2 of $n_p \leq 1000$ rpm and another value using the curve No. 2 of $n_p \leq 3000$ rpm in Fig. 19.2. Then, the pinion teeth number should be selected from the range between the two obtained values.

(b) Using the method of Fig. 19.3:

1 From the horizontal axis of Fig. 19.3, we select the rotational speed of $n_p = 1000$ rpm and go up to intersect the curve related to the helix angle of $\beta = 0$ (because the pinion is a spur type gear).
Note: If you want to employ Fig. 19.3 for the rotational speeds greater than 1000 rpm, you can consider the speed as $n_p = 1000$ rpm and then use the figure.
2 Then, we go left to obtain the minimum number of pinion teeth:

$$z_p = 24$$

Therefore, we should select a number for the pinion teeth that is greater than 24. To be on the safe side, we can select an odd or prime number (Stokes 1992) greater than 24. We choose:

$$z_p = 27$$

(c) Using the method of Fig. 19.4:
Figure 19.4 is based on equations of Feshenko (2016) (Eq. 19.5 (see Sect. 19.4)). In this method, the maximum number of pinion teeth (z_p) can be estimated based on the bending and contact strengths of pinion. It is recommended to take a number for z_p between 18 and the one obtained from Fig. 19.4.

1 First, we specify the material of pinion and the value of the coefficient M from the table in Fig. 19.4. Since the hardness of the pinion is 50 HRC, we choose the material *16MnCr5* with a hardness of 52 HRC (gas nitride) with $M = 0.0007$.
2 Then, from the right horizontal axis of Fig. 19.4, we select the speed ratio of $u = 2$ and go up to reach the curve related to spur gears (the solid line).
3 Then, we go left to intersect the curve of $M = 0.0007$.
4 Then, going down to the left horizontal axis gives us the number of pinion teeth:

$$z_p = 37$$

From the values obtained using the three methods above, we choose the maximum value, $z_p = 37$ which can lead to a smoother and quieter operation.

5 Now, we calculate the module using Eq. 19.3 (see Sect. 19.4) and round it to a standard value using Table 19.3:

$$m = \frac{D_p}{z_p} \rightarrow \mathrm{m} = \frac{150}{37} = 4.05\,\mathrm{mm}$$

According to Table 19.3, the nearest standard value is $m = 4$ mm.

6 Since we changed the calculated module of $m = 4.05$ mm to the standard value of $m = 4$ mm and selected the number of pinion teeth of $z_p = 37$, we have to recalculate the pitch diameter of pinion (D_p).

Therefore, the summary of the results is:
The number of pinion teeth: $z_p = 37$
The module: $m = 4$ mm
The pitch circle diameter of pinion, using Eq. 19.3 (see Sect. 19.4):

$$D_p = mz_p = 4 \times 37 = 148\,\mathrm{mm}$$

The number of gear teeth, using Eq. 19.4 (see Sect. 19.4):

$$z_G = uz_p = 2 \times 37 = 74$$

The pitch circle diameter of gear:

$$D_G = mz_G = 4 \times 74 = 296\,\mathrm{mm}$$

19.2.3 Specifying the Gear Quality

The quality of the gear teeth is related to the speed of the gear. The graphical method illustrated in Fig. 19.5 which presents quality levels in GOST standard, can be used for specifying the gear quality. For the common gear drives, the quality level of 8 is usually considered (Feshenko 2016).

Example 19.3 Consider a gear drive that has a pinion with a pitch circle diameter of $D_p = 150$ mm. The rotational speed of the pinion is $n_p = 1000$ rpm. Specify the pinion quality.

Solution

1. In Fig. 19.5, we select the diameter of $D_p = 150$ mm from the vertical axis and go right.
2. At the same time, from the horizontal axis, we select the rotational speed of $n_p = 1000$ rpm and go up to intersect the horizontal line of previous step.
3. As shown in Fig. 19.5, the intersection point is between quality levels of 7 and 8 for spur gears and the levels of 8 and 9 for helical gears.

Therefore, if the gear is spur, the quality level should be greater than or equal to 8, and if it is helical, the level should be greater than or equal to 9.

19.2.4 Specifying the Helix Angle

In the literature, different values have been recommended for the helix angle, as shown in Table 19.4.

In addition, the minimum helix angle can be estimated from Fig. 19.7, which is based on the relation suggested by Feshenko (2016) (Table 19.4). We show in Example how the figure is applied for this purpose.

Example 19.4 If helical gears are used in Examples 19.1 and 19.2 instead of spur gears, find the helix angle.

Solution

The inputs for this example are: $z_p = 37$, $z_G = 74$, $b = 45$ mm, $m = 4$ mm, $D_G = 296$ mm, and $D_p = 148$ mm.

1 From Fig. 19.7, we select the module value of $m = 4$ mm on the horizontal axis and go up.

Table 19.4 The recommended helix angles for helical gearings

Helix angle (β)	Reference
$\beta = 8 - 16°$ $\beta \geq sin^{-1}\frac{3.5m_n}{b}$ $m_n = m\ cos\ \beta$	Feshenko (2016)
$\beta = 8 - 25°$	Decker and Kabus (2014)
$\beta = \begin{cases} 6 - 15^{\circ} & \text{(for general gearings)} \\ 15 - 25^{\circ} & \text{(for precise gearings)} \\ 30 - 40^{\circ} & \text{(for herringbone gearings)} \end{cases}$	Zhukov and Gurevich (2014)
$\beta = 10 - 20°$	Böge (2011)

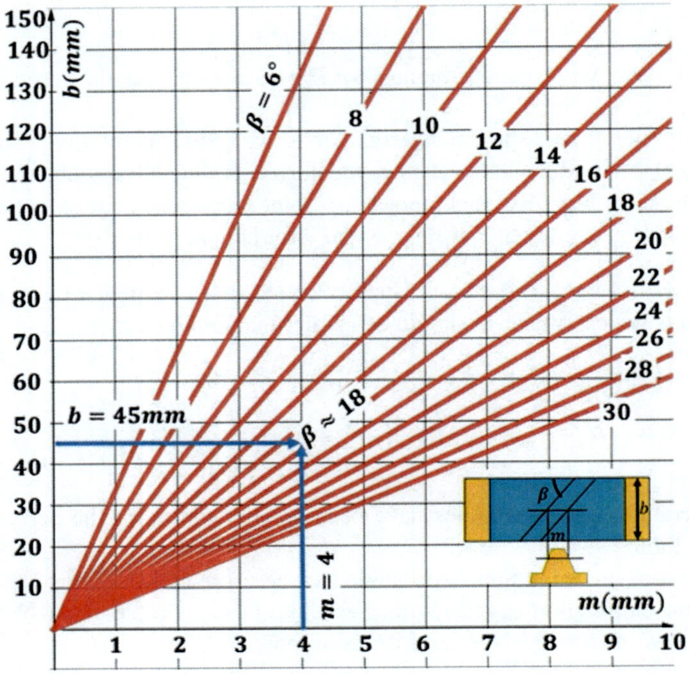

Fig. 19.7 A graph for estimating the helix angle of helical gearings

2 At the same time, we select the gear width of $b = 45$ mm from the vertical axis, and go right. The intersection point indicates the approximate value of:

$$\beta \approx 18^\circ$$

Therefore, for a good operation of the gear drive, the helix angle should be $\beta \geq 18^\circ$.

Note: The module in helical gears is the transverse module (Fig. 19.8).

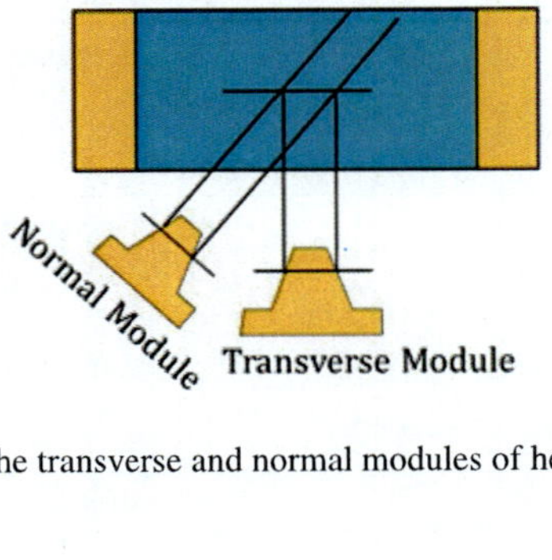

Fig. 19.8 An illustration of the transverse and normal modules of helical gearings

19.2.5 Determining the Center Distance

Some designers apply the center distance for an initial design. Figure 19.9 (based on Eq. 19.6 (see Sect. 19.4)) and Fig. 19.10 (based on Eq. 19.7 (see Sect. 19.4)) show different methods from different references that can be employed for specifying the center distance.

After determining the approximate value of the center distance, the number of pinion teeth can be obtained using Figs. 19.2, 19.3 and 19.4. Then, the module can be calculated using Eqs. 19.8, 19.9 and 19.10 (see Sect. 19.4).

Example 19.5 Find the center distance for the gear drive of Example 19.1 using Figs. 19.9 and 19.10.

Solution

The inputs for this example are: $T_1 = 750$ N m, $\varphi = 0.3$, $\sigma_H = 1000$ MPa, $z_p = 37$, and $u = 2$.

(a) Using the method of Fig. 19.9:

1. We select the torque $T = 750$ N m from the right horizontal axis of Fig. 19.9, and go up to intersect the curve of $u = 2$.
2. Then, we go left to cross both the curves related to "H_p, $H_G \geq 45$ HRC" and "$H_p \geq 45$ HRC, $H_G \leq 300$ HB". (The hardness of pinion is $H_p \geq 45$ HRC, and since we have not specified the material of gear yet, its hardness may be $H_G \leq 300$ HB or $H_G \geq 45$ HRC).

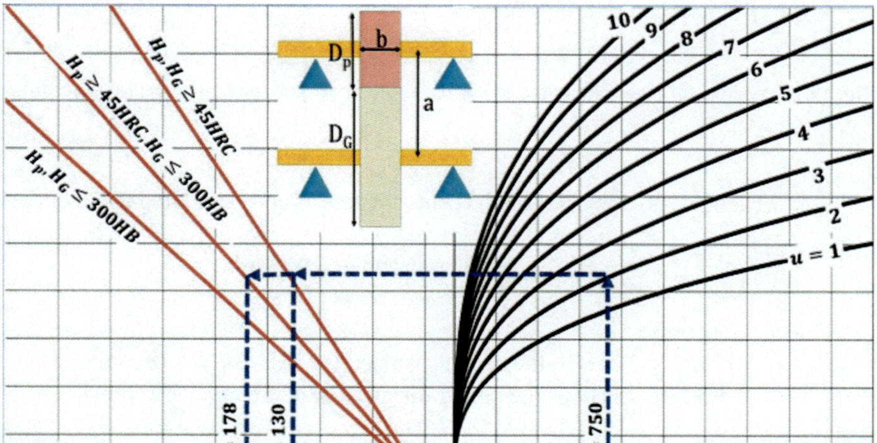

Fig. 19.9 Graphs for determining the center distance (a) of spur and helical gearings according to Eq. 19.6 (see Sect. 19.4)

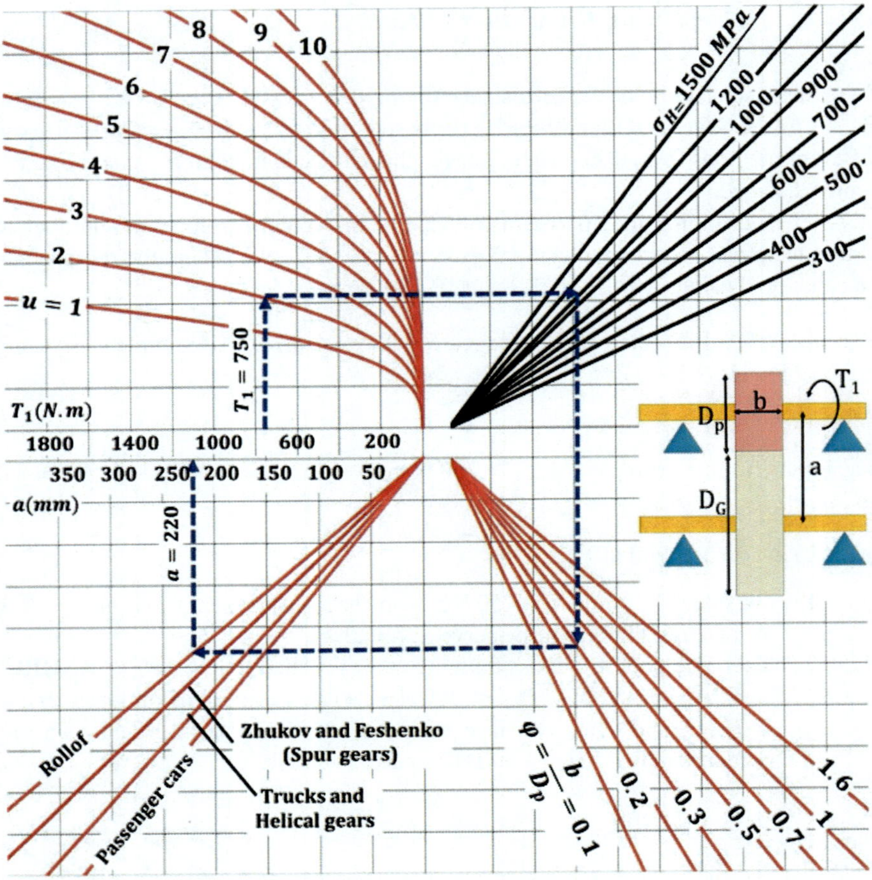

Fig. 19.10 Graphs for determining the center distance (a) of spur and helical gearings according to Eq. 19.7 (see Sect. 19.4)

3 Then, we go down to find the center distances corresponding to the curves:

$$a = 130\,\text{mm and}\, a = 178\,\text{mm}$$

Therefore:
If $H_G \leq 300$ HB, the center distance should be $a = 178$ mm.
If $H_G \geq 45$ HRC, the center distance should be $a = 130$ mm.

(b) Using the method of Fig. 19.10:

1. From the left horizontal axis of Fig. 19.10, we select the torque of $T = 750$ N m and go up to reach the curve of $u = 2$.
2. Then, we go right to cross the curve of $\sigma_H = 1000$ MPa, and go down to intersect the curve related to $\varphi = 0.3$.

3. Then, we go left to reach the curve of "Rollof". Here, we use the Rollof method (Wittel et al. 2013).
4. From that point, going up to reach the horizontal axis gives us the value of the center distance:

$$a = 220\,\text{mm}$$

5. Now, we can calculate the module, using Eqs. 19.8, 19.9 and 19.10 (see Sect. 19.4):

   i. Using Eq. 19.8:

$$\text{m} = \frac{2a}{z_p(u+1)} = \frac{2 \times 220}{37 \times (2+1)} = 3.9\,\text{mm}$$

   ii. Using Eq. 19.9:

$$\text{m} \approx (0.01 - 0.02)a \rightarrow m \approx (0.01 - 0.02) \times 220 = 2.2 - 4.4\,\text{mm}$$

   iii. Using Eq. 19.10:
   For $H_G \leq 350\,\text{HB}$:

$$\text{m} \approx (0.018)a \rightarrow m \approx 0.018 \times 220 = 3.96\,\text{mm}$$

   For $H_G \geq 45\,\text{HRC}$:

$$\text{m} \approx (0.024)a \rightarrow m \approx 0.024 \times 220 = 5.28\,\text{mm}$$

19.2.6 M.Y Method

This method relies on the assumption that the shaft diameter (d) is known. Hence, the pitch circle diameter (D_p) can be estimated simply according to the shaft diameter and using the relations presented in Table 19.5 (also in Sect. 19.4 (Eq. 19.11)).

Example 19.6 Assume that the designer has specified the pinion shaft diameter of $d = 65$ mm for the gear drive of Example 19.1. Determine the pitch circle diameter of pinion (D_p) by employing M.Y method.

Solution

1. In Example 19.1, the hardness of pinion was 50 HRC and the allowable contact stress was calculated as $\sigma_H = 1000\,\text{MPa}$.

Table 19.5 nThe relations of M.Y method for estimating the pitch circle diameter (D_p) and the gear width (b) in spur and helical gears

The pitch circle diameter (D_p):$D_p = ndD_p = nd$				
Allowable contact stress (σ_H, MPa)	$1200 - 1000$	$800 - 700$	600	500
N	$2.2 - 2.5$	3	3.5	4
The gear width (b) based on shaft arrangement				
Cantilever (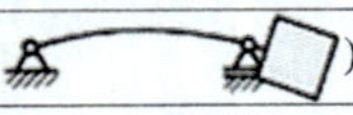):$b \geq 0.3D_p$				
Symmetrical () and asymmetrical (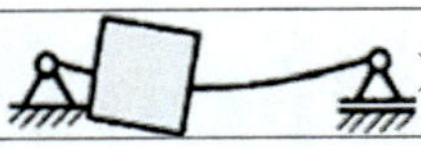):$b \geq 0.6D_p$				

2. According to Table 19.5, the value of n for $\sigma_H = 1000$ MPa is $n = 2.5$.
   Thus:
   The pitch circle diameter of pinion, using Table 19.5:

$$D_p = nd \xrightarrow{n = 2.5} D_p = 2.5 \times 65 = 165\,\text{mm}$$

   Since the arrangement of the shaft is cantilever, the pinion width, using Table 19.5 is:

$$b \geq 0.3D_p \rightarrow b \geq 0.3 \times 165 \rightarrow b \geq 49.5\,\text{mm}$$

3. If the number of pinion teeth is $z_p = 37$, the size of module will be (according to Eq. 19.3 (see Sect. 19.4)):

$$m = \frac{D_p}{zp} \xrightarrow{zp = 37} \frac{165}{37} \approx 4.5\,\text{mm}$$

19.3 Supplementary Material

19.3.1 The Material Properties of Gears

Supplemsentary Table 19.1 presents the allowable bending stress (σ_F), allowable contact stress (σ_H), and hardness for different material of gears.

Supplementary Table 19.1 the material properties of spur and helical gears

Material	Hardness Tooth side (Vickers)	Heat treatment	R_m	R_e	σ_H	σ_F	M
Grey cast iron	–	–	200	100	340	95	0.0008
Grey cast iron	–	–	250	125	350	105	0.0009
Grey cast iron	–	–	300	150	360	120	0.0009
Nodular cast iron	–	–	600	370	430	315	0.0017
Nodular cast iron	–	–	700	420	510	325	0.0012
Nodular cast iron	–	Heat treated	800	480	550	345	0.0011
Malleable cast iron	–	Normalized	500	300	380	280	0.0019
Carbon cast steel	–	Normalized	500	260	420	300	0.0017
Carbon cast steel	–	Normalized	590	300	480	336	0.0015
Carbon cast steel	600	Tooth face hard	590	300	1140	316	0.0002
36Mn5	–	Normalized	700	340	540	372	0.0013
36Mn5	–	Heat treated	750	400	560	384	0.0012
36Mn5	600	Tooth face hard	700	340	1140	352	0.0003
17CrMoV5 11	–	Normalized	650	380	520	360	0.0013
17CrMoV5 11	–	Heat treated	800	550	610	414	0.0011
30CrMoV6 4	–	Heat treated	1150	875	840	552	0.0008
ENE295	–	–	490	265	370	330	0.0024
EN S355JO	–	–	510	333	380	336	0.0023
ENE355	–	–	588	314	420	360	0.0020
ENE360	–	–	686	363	480	396	0.0017
ENC10E	650	Case-hardened	440	275	1210	500	0.0003
ENC16E	650	Case-hardened	495	295	1210	500	0.0003
ENC45	–	Normalized	540	325	430	356	0.0019
ENC45	–	Heat treated	640	390	520	410	0.0015
ENC50	600	Tooth face hard	640	390	1140	390	0.0003
ENC50	600	Face hardened	640	390	1140	605	0.0005
ENC60	–	Normalized	660	380	520	410	0.0015
ENC60	–	Heat treated	740	440	590	452	0.0013
ENC60	–	Nitro-carburized	740	440	800	650	0.0010
37MnSi5	–	Heat treated	880	635	658	493	0.0011
42MnV7	–	Heat treated	932	686	700	518	0.0011
42MnV7	550	Nitrided	800	620	930	580	0.0007
37Cr4		Heat treated	883	637	690	512	0.0011
37Cr4	600	Tooth face hard	785	539	1140	450	0.0003

(continued)

Supplementary Table 19.1 (continued)

Material	Hardness Tooth side (Vickers)	Heat treatment	R_m	R_e	σ_H	σ_F	M
37Cr4	600	Face hardened	785	539	1140	605	0.0005
37Cr4	615	Nitro-case-hard	1570	1350	1288	740	0.0004
16MnCr5	650	Case-hardened	785	588	1270	700	0.0004
35CrMo4	650	Case-hardened	880	685	1270	700	0.0004
30CrV9	800	Nitrided	800	600	1180	705	0.0005
42CrV6		Heat treated	980	850	720	530	0.0010
42CrV6	600	Tooth face hard	980	850	1160	528	0.0004
42CrV6	600	Face hardened	980	850	1160	705	0.0005
30CrMoV9	800	Nitrided	800	600	1180	705	0.0005
15NiCr6	650	Case-hardened	880	635	1270	700	0.0004
34CrNiMo6	600	Face hardened	965	750	1160	705	0.0005
34CrNiMo6	750	Nitrided	965	750	1180	730	0.0005
14NiCr14	650	Case-hardened	932	735	1270	700	0.0004
26NiCrMoV8 5	–	Heat treated	1130	980	840	602	0.0009
31NiCr14	–	Heat treated	932	785	700	518	0.0011
14NiCr18	650	Case-hardened	1130	885	1330	740	0.0004

R_m = Ultimate strength in MPa, R_e = Yield strength in MPa

19.3.2 The Module and Center Distance for Automobile Gearboxes

For automobile gearboxes, the module and center distance can be estimated using Eqs. 19.12, 19.13 and 19.14 (see Sect. 19.4).

19.3.3 The Transmission Ratio in Multi-Stage Gearboxes

In multi-stage gearboxes, there is a certain transmission ratio for each gear stage (u_1, u_2 and u_3). The overall transmission ratio (u_t) is calculated by multiplying the ratios of the gear stages. The ratios can be found in Supplementary Table 19.2, Supplementary Table 19.3 and Supplementary Fig. 19.1.

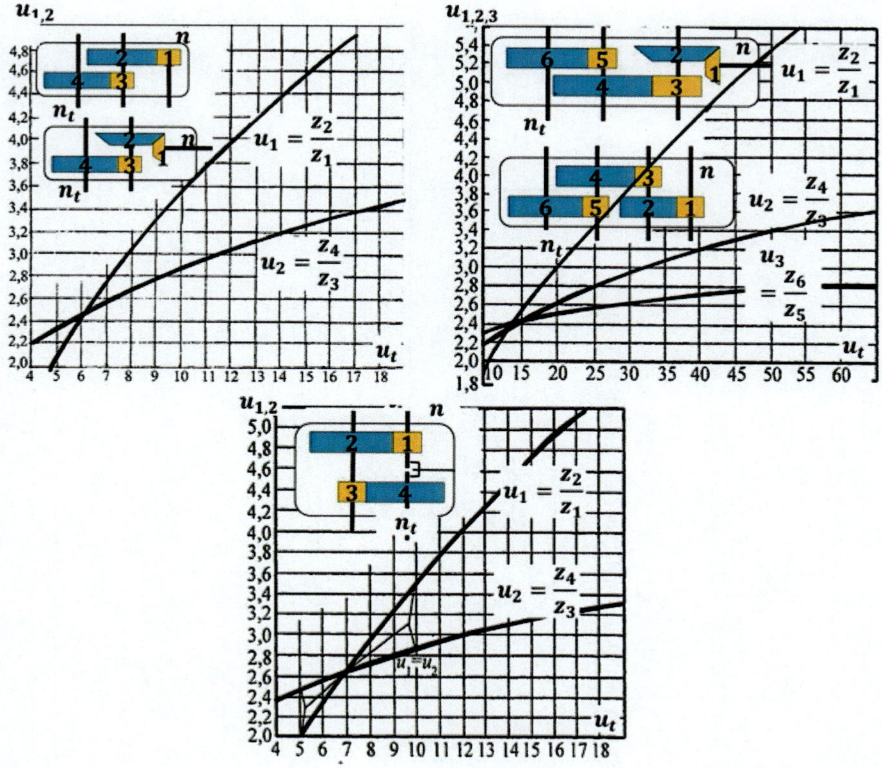

Supplementary Fig. 19.1 The transmission ratios of two- and three-stage reducers (if u_t is less than 10, for an economical manufacturing in two-stage reducers, $u_1 = u_2 = \sqrt[2]{u_t}$ is recommended) (Zhukov and Gurevich 2014)

Supplementary Table 19.2 The overall transmission ratio (u_t) and the ratios of each stage (u_1, u_2 and u_3) in multi-stage gearboxes (Decker and Kabus; Maschinenelemente 2014 Carl Hanser Verlag München)

u_t	10	20	30	40	50	70	100	200
u_1	3.5	4.8	6.5	4.1	4.5	5.5	6.9	9.5
u_2	2.9	4.2	4.6	3.3	3.5	3.9	4.5	5.5
u_3	–	–	–	2.9	3.2	3.2	3.2	3.8

Supplementary Table 19.3 The transmission ratios of the two-stage reducers (modified from Dunaev and Lelikov (2008))

GearBox arrangement	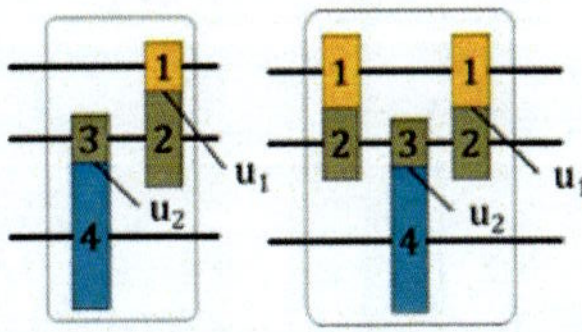	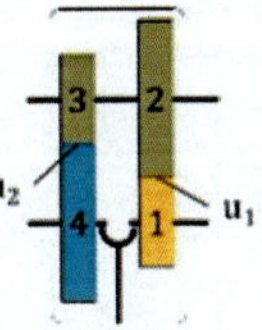	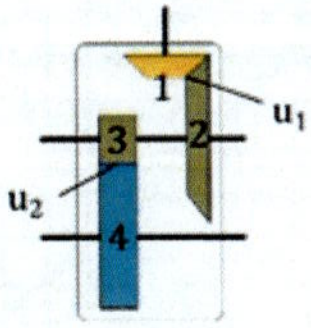	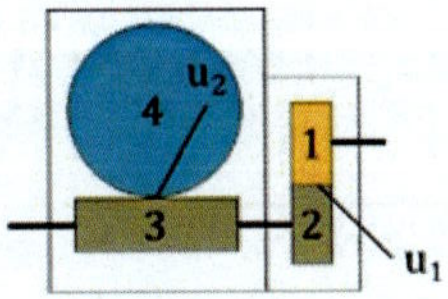
u_1	$\frac{u_t}{0.88\sqrt[2]{u_t}}$	$\frac{u_t}{0.9\sqrt[2]{u_t}}$	$\frac{u_t}{0.63\sqrt[2]{u_t}}$	1.6–3.15 or $\sqrt[5]{u_t}$
u_2	$0.88\sqrt[2]{u_t}$	$0.9\sqrt[2]{u_t}$	$0.63\sqrt[3]{u_t^2}$	$\frac{u_t}{u_1}$

19.4 Equations

The equation of Fig. 19.1 (the Roloff/Matek method (Wittel et al. 2013)):

$$D_p = 950\sqrt[3]{\frac{T_1}{\varphi\sigma_H^2} \times \frac{u+1}{u}} \quad (19.1)$$

$$\varphi = \frac{\mathrm{b}}{\mathrm{D_p}} \quad (19.2)$$

$$m = \frac{D_p}{z_p} = \frac{D_G}{z_G} \quad (19.3)$$

$$u = \frac{n_p}{n_G} = \frac{D_G}{D_p} = \frac{z_G}{z_p} \quad (19.4)$$

The equation of Fig. 19.4 (Feshenko 2016):

$$\begin{gathered} z_p = \frac{8k_0^3(1+u)}{4k_1 u} \times M \\ M = \frac{\sigma_F}{\sigma_H^2} \\ k_0 = \begin{cases} 49.5 & \text{(for spur gears)} \\ 43 & \text{(for helical gears)} \end{cases} \\ k_1 = \begin{cases} 6.8 & \text{(for spur gears)} \\ 5.8 & \text{(for helical gears)} \end{cases} \end{gathered} \quad (19.5)$$

The equation of Fig. 19.9 (Dunaev and Lelikov 2008):

$$\begin{gathered} \mathrm{a} = k_2(u+1)\sqrt[3]{\frac{T_1}{u}} \\ k_2 \approx \begin{cases} 6 & \text{(for } H_p \text{ and } H_G \geq 45\text{HRC)} \\ 8 & \text{(for } H_p \geq 45\text{HRC and } H_G \leq 300\text{HB)} \\ 10 & \text{(for } H_p \text{ and } H_G \leq 300\text{HB)} \end{cases} \end{gathered} \quad (19.6)$$

The equations of Fig. 19.10:
The method of Feshenko (2016):

$$\mathrm{a} \geq k_3(u+1)\sqrt[3]{\frac{T}{\varphi\sigma_H^2} \times \frac{u+1}{u} \times \frac{k_{H\beta}}{2}} \quad (19.7)$$

$$\mathrm{m} \geq \sqrt[3]{\frac{2k_4 T}{\varphi\sigma_F z_p^2}}$$

$$k_3 = \begin{cases} 495 & \text{(for spur gears)} \\ 430 & \text{(for helical gears)} \end{cases}$$

$$k_{H\beta} = 1.2$$

$$k_4 = \begin{cases} 68 & \text{(for spur gears)} \\ 58 & \text{(for helical gears)} \\ 52 & \text{(for herringbone gears)} \end{cases}$$

The method of Zhukov and Gurevich (2014):

$$\mathrm{a} \geq k_5(u+1)\sqrt[3]{\frac{T}{\varphi\sigma_H^2} \times \frac{u+1}{u} \times \frac{k_{H\beta}}{2}}$$

$$k_5 = \begin{cases} 480 & \text{(for spur gears)} \\ 435 & \text{(for helical gears)} \end{cases}$$

$$k_{H\beta} = 1.2$$

The method of Naunheimer et al. (2011) (for passenger cars and trucks):

$$\mathrm{a} \geq 59.84(u+1)\sqrt[3]{\frac{T}{4\varphi\sigma_H^2} \times \frac{u+1}{u} \times \frac{k_A}{2}}$$

$$k_A = \begin{cases} 0.65 & \text{(for passenger cars)} \\ 0.85 & \text{(for trucks)} \end{cases}$$

$$m = \frac{2a}{z_p(u+1)} \tag{19.8}$$

The equation for calculating the module according to Zhukov and Gurevich (2014):

$$m \approx (0.01 - 0.02)a \tag{19.9}$$

Lower values are recommended for lower speed gearings with continuous operation.

The equations for calculating the module according to Feshenko (2016):

$$\mathrm{m} \approx \begin{cases} 0.015a & \text{(for } H_p \text{ and } H_G \leq 350\text{HB)} \\ 0.018a & \text{(for } H_p \leq 45E \text{ and } H_G \leq 350\text{HB)} \\ 0.024a & \text{(for } H_p \text{ and } H_G \geq 45\text{HRC)} \end{cases} \tag{19.10}$$

The equation of M.Y method for estimating the pitch circle diameter of pinion:

$$D_p = nd$$
$$\text{n} = \begin{cases} 4 & (\text{for } \sigma_H = 500\,\text{MPa}) \\ 3.5 & (\text{for } \sigma_H = 600\ \text{MPa}) \\ 3 & (\text{for } \sigma_H = 700 - 800\ \text{MPa}) \\ 2.5 - 2.2 & (\text{for } \sigma_H = 1000 - 1200\ \text{MPa}) \end{cases} \tag{19.11}$$

The equation for calculating the center distance in automobile gearboxes (Bukharin et al. 1973):

$$a = k\sqrt[3]{T_E}$$
$$\text{k} = \begin{cases} 14.5 - 16 & (\text{for passenger cars}) \\ 17 - 19 & (\text{for trucks}) \\ 20.5 - 21.5 & (\text{for diesel cars}) \end{cases} \tag{19.12}$$

The module and gear width in automobile gearboxes (Vakhlamov 2007):

$$\text{m} = \begin{cases} 2.25 & (\text{for } TE = 100\ \text{N.m}) \\ 2.5 & (\text{for } TE = 200\ \text{N.m}) \\ 3.75 & (\text{for } TE = 400\ \text{N.m}) \\ 4.25 & (\text{for } TE = 600\ \text{N.m}) \\ 4.5 & (\text{for } TE = 800\ \text{N.m}) \\ 6 & (\text{for } TE = 1000\ \text{N.m}) \end{cases}$$
$$\text{b} = (5 - 8)m \tag{19.13}$$

The module and gear width in automobile gearboxes (Mott et al. 2018):

$$\text{b} = (8 - 16)m \approx 12m \tag{19.14}$$

References

Böge A (2011) Handbuch Maschinenbau: Grundlagen und Anwendungen der Maschinenbau-Technik. Vieweg+Teubner, Wiesbaden (in German)

Bukharin NA, Prozorov VS, Shchukin MM (1973) Automobiles. Mechanical Engineering, Moscow (in Russian)

Collins J, Busby H, Staab G (2009) Mechanical design of machine elements and machines: a failure prevention perspective, 2nd edn. Wiley, New York

Decker K-H, Kabus K (2014) Decker Maschinenelemente Funktion, Gestaltung und Berechnung. Carl Hanser Verlag, München (in German)

Dunaev PF, Lelikov OP (2008) Designing of units and machine parts. Academy, Moscow (in Russian)
Feshenko VN (2016) Designer's handbook. Book 1. Machines and mechanisms: textbook-practical. Infra-Engineering, Moscow-Vologda (in Russian)
Jiang W (2019) Analysis and design of machine elements, 1st edn. Wiley, New Jersey
Juvinall R, Marshek K (2011) Fundamentals of machine component design, 5th edn. Wiley, New York
Mott RL, Vavrek EM, Wang J (2018) Machine elements in mechanical design, 6th edn. Pearson Education, New York
Naunheimer H, Bertsche B, Ryborz J, Novak W (2011) Automotive transmissions. Springer, Berlin, Heidelberg
Stokes A (1992) Manual gearbox design. Butterworth-Heinemann, Oxford
Vakhlamov VK (2007) Design, calculation and operational properties car. Academy, Moscow (in Russian)
Wittel H, Muhs D, Jannasch D, Voßiek J (2013) Roloff/Matek Maschinenelemente Normung, Berechnung, Gestaltung. Springer Vieweg, Wiesbaden (in German)
Zhukov KP, Gurevich YE (2014) Design of parts and assemblies of machines: a textbook for universities. Mechanical Engineering, Moscow (in Russian)

Chapter 20
Bevel Gear Drives

Abbreviations

b	Teeth face width (gear width), mm
d	Shaft diameter, mm
d_e	Outer pitch diameter, mm
d_{e1}	Outer pitch diameter of pinion, mm
d_m	Mean pitch diameter, mm
d_{m1}	Mean pitch diameter of pinion, mm
H_G	Gear hardness
H_p	Pinion hardness
HB	Brinell hardness scale
HRB	Rockwell hardness scale B
HRC	Rockwell hardness scale C
HV	Vickers hardness scale
$K, k_2, k_{be}, k_{sh}, v_H, \theta_H$	Coefficients for the equations
m_e	Outer module, mm
m_m	Mean normal module, mm
n_1	Rotational speed of pinion, rpm
R_e	Outer cone distance, mm
T_1	Torque on the shaft of pinion, N.m
u	Speed ratio, transmission ratio ($u \geq 1$)
z_1	Number of pinion teeth
β	Helix angle, degree
δ_1	Pitch cone angle, degree
σ_F	Allowable bending stress, MPa
σ_H	Allowable contact stress, MPa

M. Yaghoubi and H. Tavakoli, *Mechanical Design of Machine Elements by Graphical Methods*, Materials Forming, Machining and Tribology,
https://doi.org/10.1007/978-3-031-04329-1_20

20.1 Introduction

Bevel gear drives are used to transmit or transfer power or motion between intersecting shafts. The teeth of bevel gears are machined on conical surfaces. There are two primary types for the teeth of bevel gears, namely, straight and spiral. Straight and spiral bevel gears are related to each other in the same way that spur and helical gears do. Straight bevel gears are regarded as the simplest type of gearings for intersecting shaft applications. On the other hand, spiral gears have the advantage of gradual engagement that enables them to operate more smoothly than straight bevel gears (Collins et al. 2009; Juvinall and Marshek 2011; Jiang 2019).

In this chapter, we propose methods for designing straight and spiral bevel gears in a simple way.

20.2 Design Methods

The design of bevel gears is highly dependent on the speed ratio. In addition, for determining the module, both the bending and contact strengths of pinion must be considered.

Since the small gear (pinion) is more vulnerable to failure, we base the design on it.

In this chapter, the subscripts *1* and *2* refer to the pinion and gear, respectively, as shown in Fig. 20.1.

20.2.1 The Pinion Design

The following steps indicate the process of designing bevel gears (pinions):

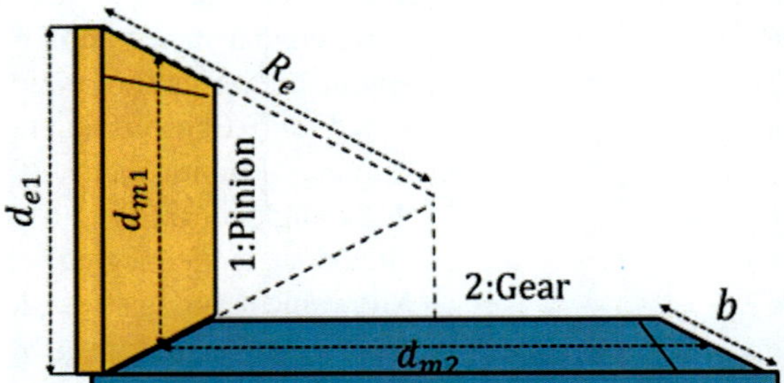

Fig. 20.1 A schematic of a bevel gear drive

Table 20.1 The allowable contact stress (σ_H) for material used in bevel gears (Dunaev and Lelikov 2008)

Heat treatment	Hardness	Material type	σ_H (MPa)
Normalized or tempered	<350 HRB	Carbon and alloy steel	2 × HB + 70
Induction hardening	38–56 HRC		17×HRC+200
Cemented or carburized	>56 HRC	Alloy steel	23 × HRC
Nitride	550–750 HV		1050–1150
Without thermal processing	–	Cast iron	400–500

1. First, according to the material and hardness of the pinion, obtain the allowable contact stress (σ_H) and allowable bending stress (σ_F) from Tables 20.1 and 20.2.
2. Then, determine an initial estimate for the mean pitch diameter of pinion (d_{m1}) using Fig. 20.2.
3. Then, specify the number of pinion teeth (z_1) from Figs. 20.3 and 20.4.
4. Then, find the mean module (m_m) from Figs. 20.2 and 20.5. Between the two obtained values, select the bigger one and round it to a standard module according to Table 20.3.
5. Then, calculate the mean pitch diameter of pinion (d_{m1}) using Eq. 1.1 (see Sect. 20.3).
6. Then, determine the pinion (or gear) width (b) using Fig. 20.5.
7. The gear teeth quality depends on the velocity of the pinion. The higher the velocity, the higher the quality should be. Specify the quality of the gears using Fig. 20.6.

Example 20.1 A straight bevel gear drive is schematically shown in Fig. 20.7. The rotational speed (n_1) and torque (T_1) of the pinion are 800 rpm and 400 N.m, respectively. The transmission (or speed) ratio is $u = 2$. Assume that the pinion material is 42CrMo4 with bending strength of $\sigma_F = 770$ MPa and contact strength of $\sigma_H = 1070$ MPa (Table 20.2). Determine the mean module (m_n) and the number of pinion teeth (z_1).

Solution

First, we should estimate the mean module according to the contact and bending strengths, and then choose the bigger one and round it to a standard value.

(a) Estimating the mean module based on the contact strength:

1. First, we should find the value of coefficient K from Fig. 20.2 (from the curves in the third quadrant (the lower left corner)). Two methods are presented in the figure for this purpose, "*Rollof*" (Wittel et al. 2013) and "*Zhukov*" (Zhukov and Gurevich 2014). We arbitrarily choose the method of "*Zhukov*".

Table 20.2 The material properties of bevel gears (Decker, Kabus; Maschinenelemente, 2014 Carl Hanser Verlag München)

Material	σ_F	σ_H	Hardness	Heat treatment of steel
EN-GJL-200 (GG-20)	80	300	180 HB	–
EN-GJL-250 (GG-25)	110	360	220 HB	–
EN-GJMB-350 (GTS-35)	330	320	150 HB	–
EN-GJMB-650 (GTS-65)	410	460	220 HB	–
EN-GJS-400 (GGG-40)	370	370	180 HB	–
EN-GJS-600 (GGG-60)	450	490	250 HB	–
EN-GJS-800 (GGG-80)	500	600	320 HB	–
GS-52	280	320	160 HB	–
GS-60	320	380	180 HB	–
E295 (St 50)	320	370	160 HB	–
E335 (St 60)	350	430	190 HB	–
E360 (St 70)	410	460	210 HB	–
C 45	410	530	190 HV 10	Normalized
34CrMo4	520	530	270 HV 10	Hardened and tempered
42CrMo4	570	600	300 HV 10	Hardened and tempered
34CrNiMo6	610	630	310 HV 10	Hardened and tempered
16MnCr5	860	1470	720 HV 10	Case hardened
15CrNi6	920	1490	730 HV 10	Case hardened
17CrNiMo6	1000	1510	740 HV 10	Case hardened
42CrMo4	770	1070	550 HV 10	Gas nitride
16MnCr5	810	1100	550 HV 10	Gas nitride
31CrMoV9	840	1230	700 HV 10	Gas nitride
C 45	620	710	420 HV 10	Nitrocarburized
16MnCr5	650	770	560 HV 10	Nitrocarburized
42CrMo4	680	830	610 HV 10	Nitrocarburized
34Cr4	900	1350	650 HV 10	Carbonitrided

2. From the horizontal axis of Fig. 20.2 (the third quadrant (the lower left corner)), we select the speed ratio of $u = 2$, and go up to intersect the curve of "*Straight*" (the dashed line). Then, we go left to obtain the value of K:

$$K = 1200$$

3. Now, in the second quadrant (the upper left corner) of Fig. 20.2, we select the torque of $T_l = 400$ N.m from the horizontal axis, and go up to reach the curve related to $K = 1200$.

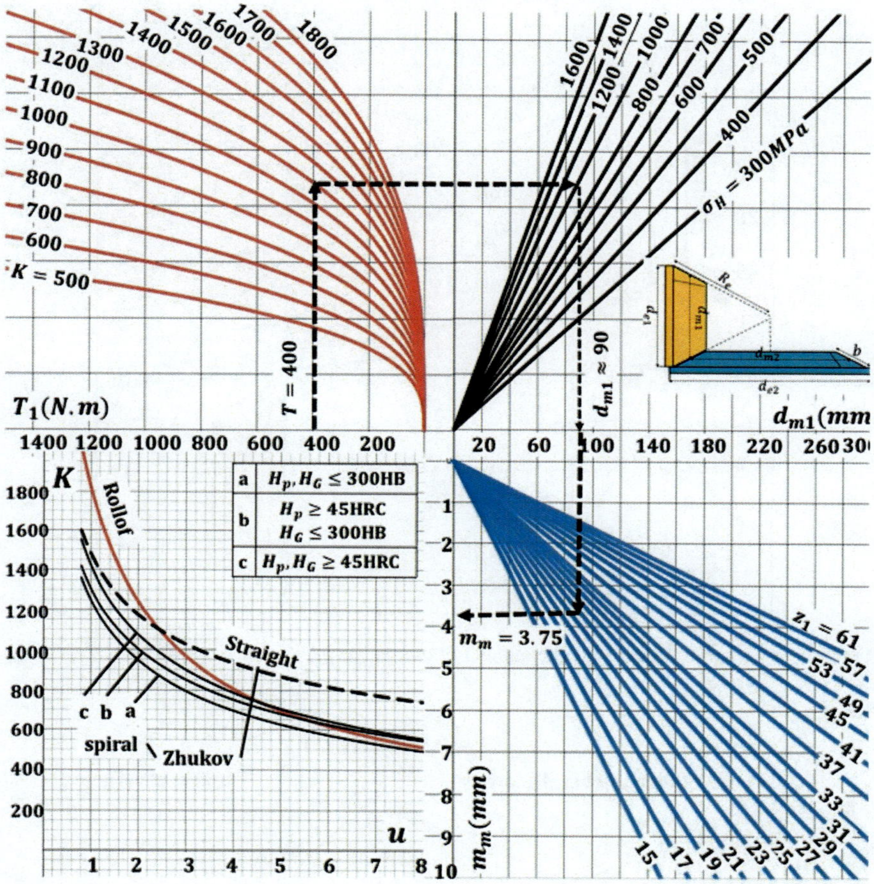

Fig. 20.2 Graphs for estimating the mean pitch diameter (d_{m1}), and mean module (m_m) of bevel gears based on the contact strength (σ_H)

4. Then, we go right to cross the curve of $\sigma_H = 1070$ MPa. From the crossed point, we go down to achieve the mean pitch diameter of pinion (d_{m1}):

$$d_{ml} = 90\,\text{mm}$$

To continue the estimation of the mean module in Fig. 20.2, we need to have the number of pinion teeth (z_1). We can find a suggested value for z_1 from Figs. 20.3 and 20.4.

(Unfortunately, there are different points of view in this regards in the literature.)

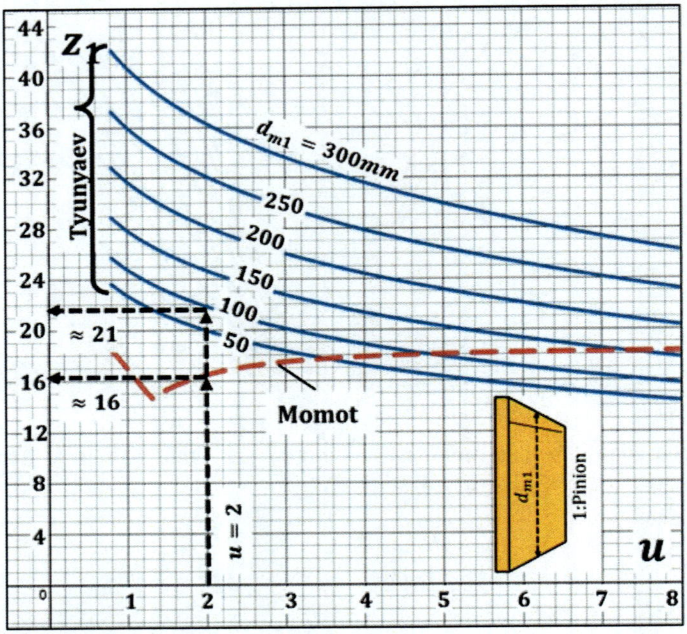

Fig. 20.3 A graph for specifying the number of pinion teeth (z_1) in bevel gears

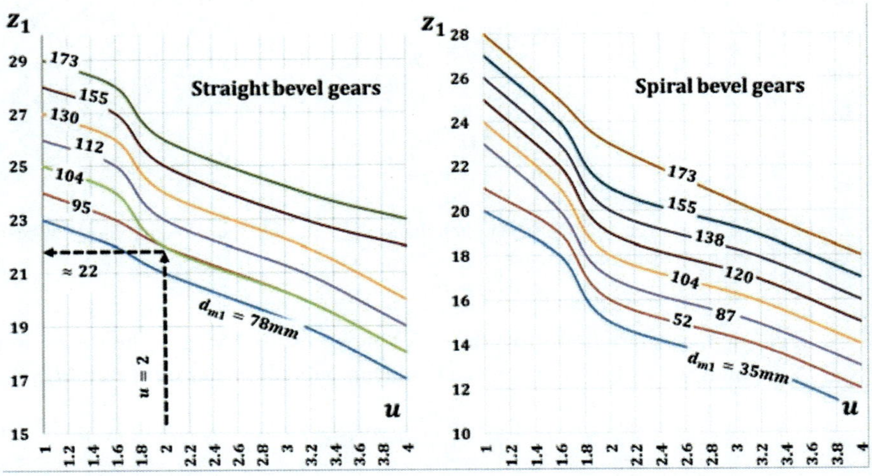

Fig. 20.4 Graphs for specifying the number of pinion teeth (z_1) in straight and spiral bevel gears (multiply the values by 1.6 for H_p and $H_G \leq 350$ HB and by 1.3 for $H_p > 45$ HRC and $H_G \leq 350$ HB)

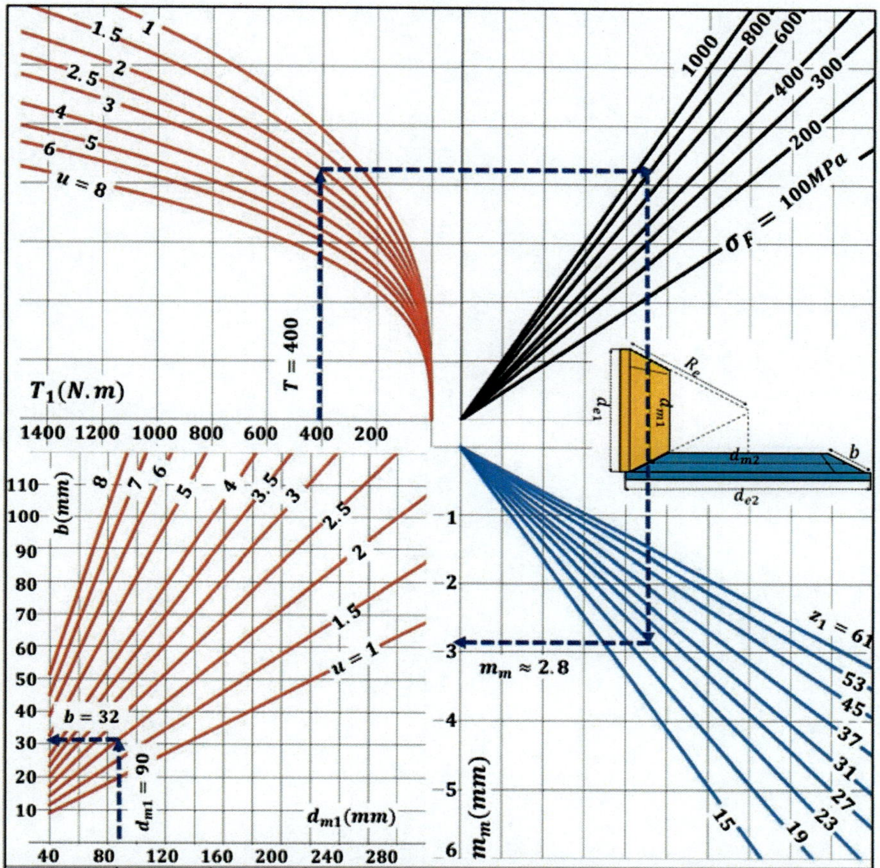

Material	GG20	GG25	GTS35	GTS65	GG40	GG60	GGG80	GS52
Hardness	180HB	220HB	150HB	220HB	180HB	250HB	320HB	160HB
σ_F	80	110	330	410	370	450	500	280
GS60	E295	E335	E360	C 45	34CrMo4	42CrMo4	34CrNiMo6	16MnCr5
180HB	160HB	190HB	210HB	190HV10	270HV10	300 HV 10	310 HV 10	720 HV 10
320	320	350	410	410	520	570	610	860
15CrNi6	17CrNiMo6	42CrMo4	16MnCr5	31CrMoV9	C 45	16MnCr5	42CrMo4	34Cr4
730 HV 10	740 HV 10	550 HV 10	550 HV 10	700 HV 10	420 HV 10	560 HV 10	610 HV 10	650 HV 10
920	1000	770	810	840	620	650	680	900

Fig. 20.5 Graphs for estimating the mean module (m_m) of bevel gears based on the bending strength (σ_F)

5. We select the speed ratio of $u = 2$ from the horizontal axis of Fig. 20.3 and the left horizontal axis of Fig. 20.4, and go up to intersect the approximate curves related to $d_{mI} = 90$ mm.
6. Then, in the both figures, we go left to find the suggested numbers of pinion teeth (z_I):

Table 20.3 Standard modules of bevel gears (all values are in mm)

Series 1	0.05	0.06	0.08	0.1	0.12	0.16	0.2	0.25	0.3	0.4	0.5	0.6	0.7	0.8	0.9	1	1.25
	1.5	2	2.5	3	4	5	6	8	10	12	16	20	25	32	40	50	60
Series 2	0.055	0.07	0.09	0.11	0.14	0.18	0.22	0.28	0.35	0.45	0.55	0.65	0.75	0.85	0.95	1.125	1.375
	1.175	2.25	2.75	3.5	4.5	5.5	7	9	11	14	18	22	28	36	45	55	70

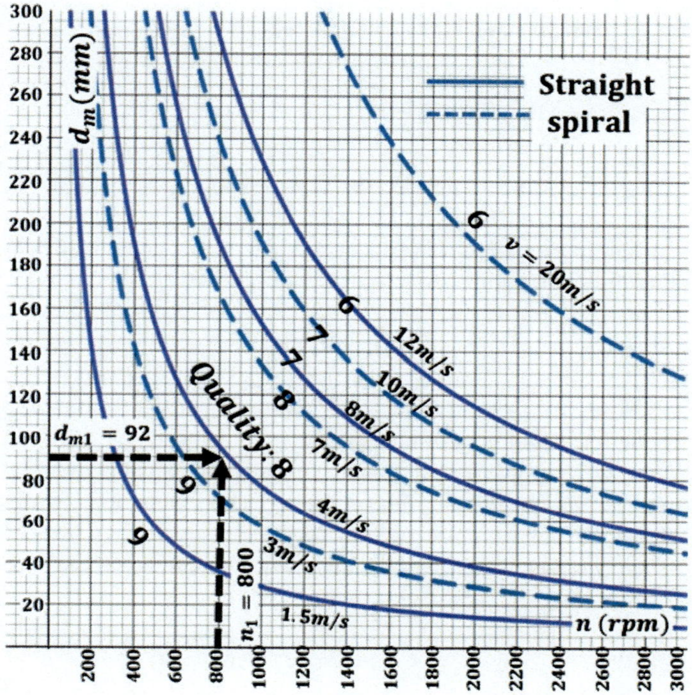

Fig. 20.6 A graph for specifying the gear quality of bevel gearings in GOST standard

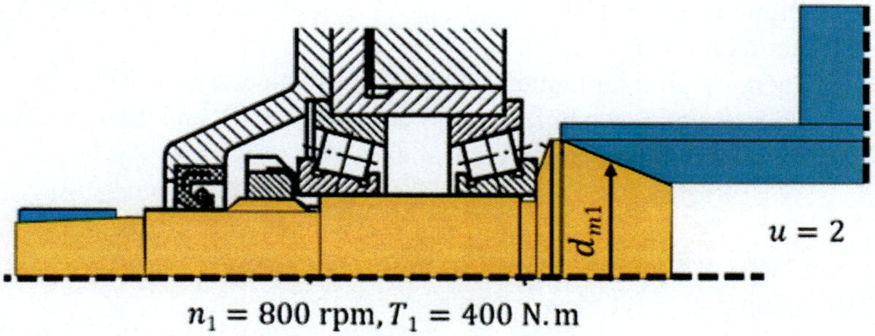

Fig. 20.7 An illustration for the bevel gear drive of Example 20.1

From the curve of "*Momot*" (Momot and Yanchevsky 2011) in Fig. 20.3: $z_1 = 16$.

From the curve of "*Tyunyaev*" (Tyunyaev et al. 2013) in Fig. 20.3: $z_1 = 21$.

From Fig. 20.4: $z_1 = 22$.

However, in design, it is recommended to choose $z_I = 18 - 30$, (for straight teeth bevel gears: $z_1 \geq 25 \cos \delta_1$, and for spiral teeth bevel gears:

$z_1 \geq 15 \cos \delta_1 \cos \beta$) (Feshenko 2016). Thus:

$$\delta_1 = \tan^{-1}\left(\frac{1}{u}\right) \xrightarrow{u=2} \delta_1 = 26.56^\circ$$

$$z_1 \geq 25 \cos \delta_1 \rightarrow z_1 \geq 25 \cos 26.56^\circ \rightarrow z_1 \geq 25 \times 0.89 \rightarrow z_1 \geq 22.3$$

It is preferred to choose an odd or prime number (Stokes 1992), therefore, the number of pinion teeth is chosen as:

$$z_1 = 23$$

Notes: the method or curve of "*Momot*" gives the minimum suggested value for the number of pinion teeth.

7. Now, we continue determining the mean module from Fig. 20.2. From the point of $d_{ml} = 90$ mm mm on the right horizontal axis, we keep going down to reach the curve of $z_I = 23$.
8. Then, we go left to obtain the mean normal module on the vertical axis:

$$m_{ms} = 3.75\,\text{mm}$$

(b) Estimating the mean module based on the bending strength:

1. In the second quadrant (the upper left corner) of Fig. 20.5, we find the torque of $T_I = 400$ N.m on the horizontal axis, and go up to cross the curve of $u = 2$.
2. Then, we go right to intersect the curve related to $\sigma_F = 770$ MPa.
3. Then, we go down to reach the curve of $z_1 = 23$, and then go left to find the mean module on the vertical axis:

$$m_m = 2.8\,\text{mm}$$

(c) Determining the final mean module:

From the two obtained values for the mean module, we choose the higher one, $m_m = 3.75$ mm. Afterwards, we should round it to a standard size from Table 20.3.

Some engineers use the outer module (m_e) as the standard module, which is equal to $m_e \approx 1.16 m_m$ (it should be rounded). Therefore:

According to the mean pitch module:

$$m_m = 3.75\,mm \xrightarrow[\text{}]{\text{round to a standard value}} mm = 4\,mm$$

According to the outer pitch module:

$$m_e = 1.16 m_m = 1.16 \times 3.75 = 4.35\,mm \xrightarrow[\text{standard value}]{\text{round to a}} m_e = 4.5\,mm$$

(d) Estimating the gear (or pinion) width:

To obtain the gear width, we can use the graph in the third quadrant (the lower left corner) of Fig. 20.5.

1. From the horizontal axis, we select $d_{mI} = 90$ mm, and go up to intersect the curve of $u = 2$.
2. Then, we go left to achieve the gear width:

$$b = 32\,\text{mm}$$

In the end, a summary of the result is as follows:
The mean module: $m_m = 4$ mm.
The number of pinion teeth: $z_I = 23$.
The gear width: $b = 32$ mm.
The mean pitch diameter (since we rounded the mean module to a standard value, we have to recalculate the mean pitch diameter using Eq. 1.1 (see Sect. 20.3)):

$$d_{m1} = m_m \times z_1 = 4 \times 23 = 92\,\text{mm}$$

Note: according to the results, the mean pitch diameter should be $d_{m1} \geq 92$ mm. If you have to increase the mean pitch diameter due to design or manufacturing considerations, increase the number of teeth or the size of module. Never reduce the module size, otherwise you will have to repeat the steps above.

20.2.2 Specifying the Gear Quality

The quality of gear teeth is highly related to the velocity of gear. To specify the quality level in GOST standard, Fig. 20.6 can be employed. For the common bevel gear drives, the quality level is typically considered to be 8 (Feshenko 2016).

Example 20.2 specify the quality of the gear teeth for the bevel gearings of Example 20.1. The rotational speed of pinion was $n_I = 800$ rpm and the final size for the mean pitch diameter of the pinion obtained $d_{mI} = 92$ mm.

Solution

1. In Fig. 20.6, we select the mean pitch diameter of $d_{m1} = 92$ mm from the vertical axis and go right.
2. At the same time, we select the rotational speed of $n_1 = 800$ rpm from the horizontal axis, and go up to intersect the horizontal line of previous step.

As shown in Fig. 20.6, if the bevel gear is spiral, the gear quality level should be more than 9 and if it is straight, the quality should be approximately 8.

20.2.3 M.Y Method I

In terms of manufacturability, for gear drives in which the shaft and pinion are one-piece (Fig. 20.8a), the mean pitch diameter should be $d_{m1} \geq 1.25d_1$; and for the ones in which the shaft and pinion are connected by keys (Fig. 20.8b), it should be $d_{m1} \geq 2.5d_1$ (Wittel et al. 2013). The shaft diameter, d_1, is usually considered to be $d_1 \approx d + (5 - 20)$mm.

In M.Y method I, the pinion can be designed both according to the contact and bending strengths and also evaluated in terms of manufacturability. Because the mean pitch diameter (d_{m1}) should not be less than the shaft diameters, d_1 and d (Fig. 20.8).

In Fig. 20.9 you can find shaft diameter (d) and the mean pitch diameter of the pinion (d_{m1}) easily.

Example 20.3 Assume that the shaft and pinion of a bevel gear drive are connected by a rectangular key, similar to Fig. 20.8b. The torque on the pinion and the speed ratio are $T_1 = 400$ N.m and $u = 2$, respectively. The material of the pinion is 42CrMo4 with the contact strength of $\sigma_H = 1070$ MPa. Estimate the mean module (m_m) and the mean pitch diameter (d_{m1}) of the pinion by M.Y method I.

Solution

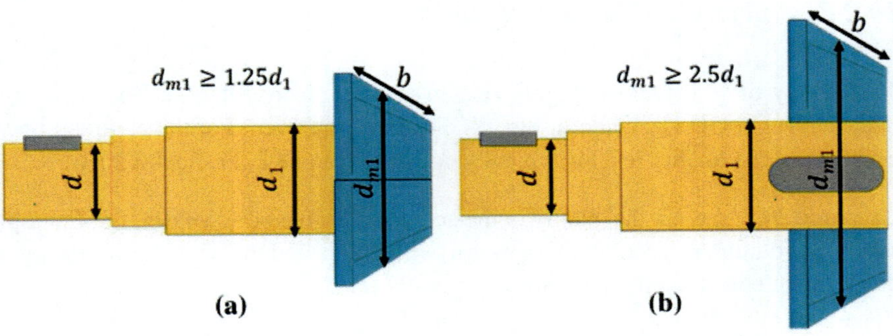

Fig. 20.8 An illustration for the connection of the shaft and pinion in bevel gears, (**a**) the one-piece design (**b**) the connection by keys

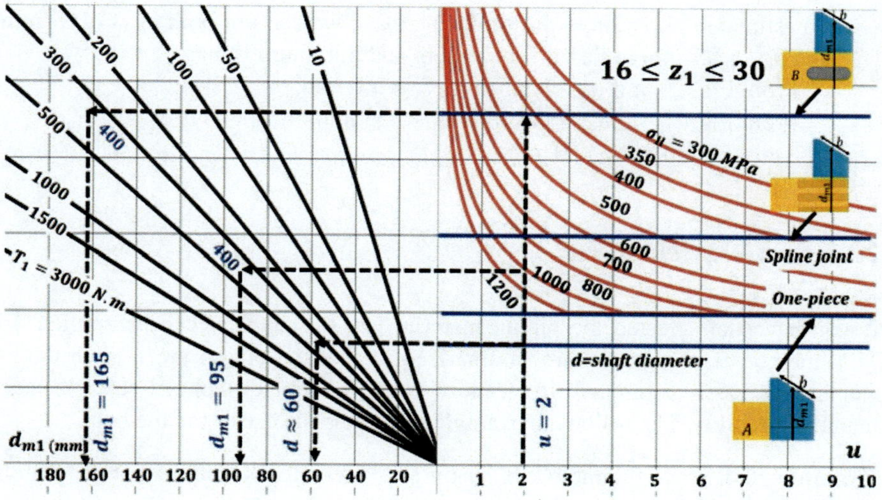

Fig. 20.9 Graphs for determining the shaft diameter (d) and the mean pitch diameter of the pinion (d_{m1}) in M.Y method I

1. In Fig. 20.9, from the right horizontal axis, we select $u = 2$, and go up to reach the curve of $\sigma_H = 1070$ MPa and the curve related to the pinion which is connected to the shaft by a rectangular key.
2. Then, we go left to cross the approximate curve of $T_1 = 400$ N.m, and go down to find mean pitch diameter of the pinion:
   In terms of the manufacturability: $d_{m1} = 165$ mm.
   In terms of the strength: $d_{m1} = 95$ mm
3. Between the two values, we select the larger one, i.e., $d_{m1} = 165$ mm.
4. In Fig. 20.9, we can also find the shaft diameter (d):

$$d = 60\,\text{mm}$$

5. We select the number of the pinion teeth from the range of $16 \leq z_1 \leq 30$. Arbitrarily we choose:

$$z_1 = 23$$

6. Then, we calculate the mean module using Eq. 1.1 (see Sect. 20.3):

$$m_m = \frac{d_{m1}}{z_1} = \frac{165}{23} = 6.9 \xrightarrow{\text{round to a standard value}} m_m = 7mm$$

1. To estimate the pinion width from Fig. 20.5, we can use both $d_{m1} = 165$ mm and $d_{m1} = 95$ mm to find the maximum and minimum values:
   From Fig. 20.5, if $d_{m1} = 165$ mm: $b \approx 60$ mm.
   From Fig. 20.5 if $d_{m1} = 95$ mm: b $\approx$ 35 mm.
   Thus: $35\text{mm} \leq b \leq 60\,\text{mm}$.

20.2.4 M.Y Method II

If you have not selected the pinion material yet, this method can be applied. In M.Y method II, the graphs are summarized so that the pinion mean pitch diameter can be easily estimated, in terms of bending strength, contact strength, and manufacturability. The following example illustrates the use of the method.

Example 20.4 Solve Example 20.3 using M.Y method II. Assume that the pinion material has not been specified.

Solution

1. From the horizontal axis of Fig. 20.10, we select the torque of, and go up to reach the curve of $u = 2$ and the range or area B (the area B corresponds to the

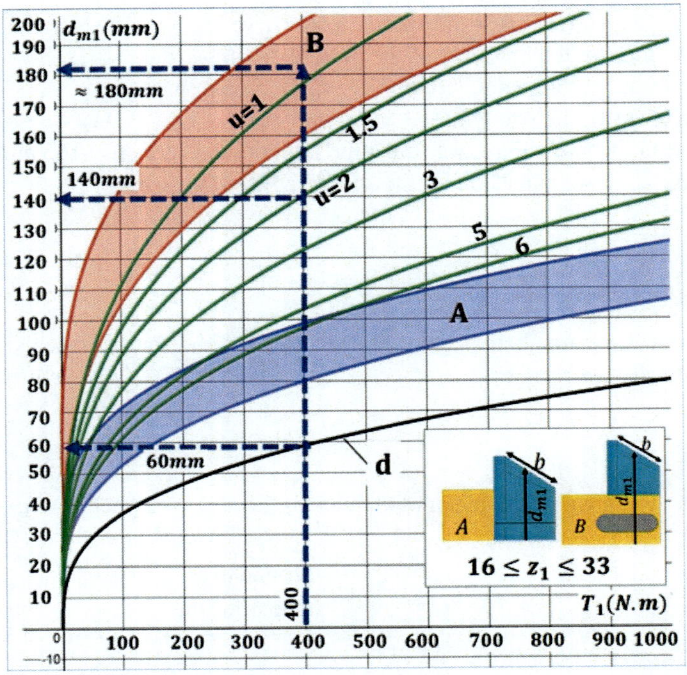

Fig. 20.10 A graph for estimating the pinion mean pitch diameter (d_{m1}) using M.Y method II

pinions that are connected to the shaft by a rectangular key). In the area B, we arbitrarily choose the middle of the range.

2. Then, we go left to find the mean pitch diameter of the pinion:
   In term of durability: $d_{ml} \geq 140\,\text{mm}$.
   In term of manufacturability:$d_{m1} \geq 180\,\text{mm}$
3. We select the maximum value, $d_{ml} = 180\,\text{mm}$ as the mean pitch diameter of the pinion.
   We can also find the minimum shaft diameter (d) from the curve of "d" in Fig. 20.10, which is:

$$d = 60\ \text{mm}$$

4. We select the number of the pinion teeth (z_I) in the range of $16 \leq z_1 \leq 33$. We arbitrarily take:

$$z_1 = 23$$

5. We can also find the number of pinion teeth from Figs. 20.3 and 20.4.
   Note: the fewer the teeth number, the stronger the teeth (in terms of bending strength).
6. Then, we calculate the mean module using Eq. 1.1 (see Sect. 20.3):

$$m_m = \frac{d_{m1}}{z_1} = \frac{180}{23} = 7.8 \xrightarrow[\text{}]{\text{round to a standard value}} m_m = 8mm$$

7. The maximum and minimum values of the pinion width (b) from Fig. 20.5 with $d_{ml} = 180$ mm and $d_{ml} = 140$ mm is:

$$50\text{mm} \leq b \leq 65\,\text{mm}$$

20.3 Equations

$$m_m = \frac{d_{m1}}{z_1} \tag{20.1}$$

The equation of M.Y method I for estimating the mean pitch diameter (d_{ml}) (Fig. 20.9):

$$d_{m1} = \max\begin{cases} 16z_1 \times \sqrt[3]{\dfrac{1.3\times 4T_1}{0.275z_1^2\sigma_F\sqrt[2]{1+u^2}}} \\ 966\sqrt[3]{\dfrac{1.3T_1}{0.275u\sigma_H^2}} \\ (2.5\ \text{or}\ 1.25)d_1 \end{cases}$$

$$d_1 = 1.05 \times 1.18 \times d \text{ or } d + (5 - 20)\text{mm}$$

$$d = 8\sqrt[3]{T_1} \tag{20.2}$$

The equations of M.Y method II for estimating the mean pitch diameter (d_{m1}) and the shaft diameter (d) (Fig. 20.10):

$$d_{m1} = \frac{26}{1.16}\sqrt[3]{\frac{T_1}{0.85u}}$$

$$d = 8\sqrt[3]{T_1} \tag{20.3}$$

The equation of the "*Rollof*" method (Wittel et al. 2013) in Fig. 20.2:

$$d_{m1} = K\sqrt[3]{\frac{T_1}{\sigma_H^2}}$$

$$K = 205\sqrt[3]{\frac{\sin(\delta_1)}{u}} \tag{20.4}$$

The equation of the "*Zhukov*" method (Zhukov and Gurevich 2014) in Fig. 20.2:

$$d_{e1} = K\sqrt[3]{\frac{T_1}{\sigma_H^2}}$$

$$K = k_d\sqrt[3]{\frac{1}{\theta_H u(1 - 0.5k_{be})^2 k_{be} u}}$$

$$k_d = \begin{cases} 960 \text{ (for straight bevel gears)} \\ 860 \text{ (for spiral bevel gears)} \end{cases}$$

θ_H for straight bevel gears:

$$\theta_H = 0.85$$

θ_H for spiral bevel gears:

$$\theta_H = \begin{cases} 0.87 + 0.15u \ (for\ H_p \text{ and } H_G \geq 45HRC) \\ 0.8 + 0.092u \ (for\ H_p \geq 45HRC \text{ and } H_G \leq 300HB) \\ 0.58 + 0.11u \ (for\ H_p \text{ and } H_G \leq 300HB) \end{cases} \tag{20.5}$$

The equation of the "*Tyunyaev*" method (Tyunyaev et al. 2013) in Fig. 20.3:

$$z_{\min} = \sqrt[2]{(22 - 9\log(u))^2 + (6.25 - 4\log(u))\frac{d_{e1}^2}{645}} \tag{20.6}$$

The equation of the "*Momot*" method (Momot and Yanchevsky 2011) in Fig. 20.3:

$$z_{\min} = 18.4\cos\delta_1 \tag{20.7}$$

The equation of Fig. 20.5:

$$m_m = 37.5\sqrt[3]{\frac{T_1\sin(\delta_1)}{z_1^2\sigma_F}}$$

$$b = 0.15d_{e1}\sqrt{u^2+1} \tag{20.8}$$

The equations of Dunaev and Lelikov (2008):

$$d_{e1} = k_2\sqrt[3]{\frac{T_1}{u\nu_H}}$$

$$d_m = \frac{k_2 d}{8.8\sqrt[3]{u}} \approx (2.5 - 3.4)\frac{d}{\sqrt[3]{u}}$$

ν_H for straight bevel gears:

$$\nu_H = 0.85$$

ν_{H} for spiral bevel gears:

$$\nu_H = \begin{cases} 1.22 + 0.21u\ (for\ H_p \text{ and } H_G \geq 45HRC) \\ 1.13 + 0.13u\ (for\ H_p \geq 45HRC \text{ and } H_G \leq 300HB) \\ 0.81 + 0.15u\ (for\ H_p \text{ and } H_G \leq 300HB) \end{cases}$$

$$k_2 = \begin{cases} 22\ (for\ H_p \text{ and } H_G \geq 45HRC) \\ 25\ (for\ H_p \geq 45HRC \text{ and } H_G \leq 300HB) \\ 30\ (for\ H_p \text{ and } H_G \leq 300HB) \end{cases} \tag{20.9}$$

The equations of Momot and Yanchevsky (2011):

$$m_m = (0.01 - 0.03)ud_{m1}$$

$$z_1 = \frac{1}{u(0.01 - 0.03)}$$

(Smaller values are for larger gear ratios and vice versa.)

$$\delta_1 = \tan^{-1}\left(\frac{1}{u}\right)$$

$$d_{e1} = \frac{d_{m1}}{(1 - 0.5k_{be})} \approx 1.16 d_{m1}$$

$$k_{be} = \frac{b}{R_e} = 0.275 \tag{20.10}$$

References

Collins J, Busby H, Staab G (2009) Mechanical design of machine elements and machines: a failure prevention perspective, 2nd edn. Wiley, New York

Decker K-H, Kabus K (2014) Decker Maschinenelemente Funktion. Carl Hanser Verlag, München (in German), Gestaltung und Berechnung

Dunaev PF, Lelikov OP (2008) Designing of units and machine parts. Academy, Moscow (in Russian)

Feshenko VN (2016) Designer's handbook. Book 1. Machines and mechanisms: textbook-practical. Infra-Engineering, Moscow-Vologda (in Russian)

Jiang W (2019) Analysis and design of machine elements, 1st edn. Wiley, New Jersey

Juvinall R, Marshek K (2011) Fundamentals of machine component design, 5th edn. Wiley, New York

Momot DI, Yanchevsky IV (2011) Mechanical transmissions: Calculation of strength and design drive machines. Khnadu, Kharkiv (in Ukrainian)

Stokes A (1992) Manual gearbox design. Butterworth-Heinemann, Oxford

Tyunyaev AV, Zvezdakov VP, Wagner VA (2013) Machine parts: Textbook. Lan Electronic Library, Moscow (in Russian)

Wittel H, Muhs D, Jannasch D, Voßiek J (2013) Roloff/Matek Maschinenelemente Normung, Berechnung, Gestaltung. Springer Vieweg, Wiesbaden (in German)

Zhukov KP, Gurevich YE (2014) Design of parts and assemblies of machines: a textbook for universities. Mechanical engineering, Moscow (in Russian)

Chapter 21
Wormgear Drives

Nomenclature

a	Center distance, mm
b_1	Length of worm, mm
b_2	Face width of wormgear, mm
d_G	Pitch diameter of wormgear, mm
d_w	Pitch diameter of worm, mm
d	Shaft diameter, mm
HB	Brinell hardness scale
K_s	A material related coefficient forermining the pitch diameter of wormgears
m	Module of wormgear, mm
n_1	Rotational speed of worm, rpm
n_2	Rotational speed of wormgear, rpm
T_1	Torque on worm, N.m
T_2	Torque on wormgear, N.m
u	Gear ratio, speed ratio
v	Pitch line velocity, m/s
z_1	Number of worm threads
z_2	Number of wormgear teeth
σ_H	Allowable contact stress, MPa
φ	Ratio of pitch diameter of worm to center distance

M. Yaghoubi and H. Tavakoli, *Mechanical Design of Machine Elements by Graphical Methods*, Materials Forming, Machining and Tribology,
https://doi.org/10.1007/978-3-031-04329-1_21

21.1 Introduction

Wormgear drives, or worm gearings, are used to transmit power and motion between shafts whose axes are neither parallel nor intersecting. A worm gearset, as shown in Fig. 21.1, is composed of a worm that resembles a power screw, in mesh with a wormgear (or worm wheel) having a similar appearance as a helical gear. Worms may be single-thread or multiple-thread (Fig. 21.1). The shaft axes of the worm and wormgear are typically 90° apart. Worm gearing are often used for applications in which the speed ratios are quite high (up to $u = 360$) (Ugural 2015; Mott et al. 2018; Jiang 2019).

In this chapter, we introduce graphical methods for easy design of wormgear drives.

21.2 Design Methods

Two methods have been proposed in the literature for designing wormgear drives. Method I is based on Wittel et al. (2013) and Decker and Kabus (2014), and method II is according to Böge (2011) and Zhukov and Gurevich (2014). The following subsections provide details about the methods, as well as graphs that we employ for the design.

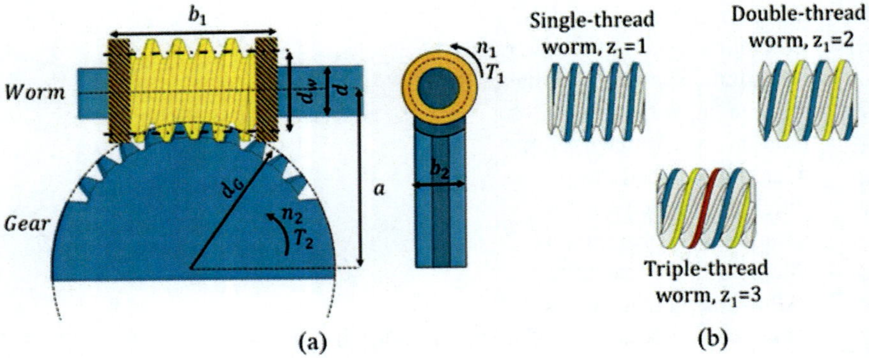

Fig. 21.1 an illustration representing geometry of a warmgear drive (**a**) and worms with different number of threads (**b**)

21.2.1 Method I

In this method, you should take the following steps:

1. First, determine the number of worm threads (z_1) based on the speed ratio (u) from Table 21.1.
2. Then, find the allowable contact stress σ_H of the wormgear and worm according to their material from Table 21.2.
3. Then, from Fig. 21.2, obtain estimated values for the center distance (a) and the pitch diameter of worm (d_w).

   In Fig. 21.2, the pitch diameter of worm (d_w) is determined based on both the allowable contact stress of the teeth and the strength of the shaft. Between the two obtained values, select the higher one.
4. Then, calculate the module (m), the pitch diameter (d_G), the number of teeth (z_2) and the width (b_2) of wormgear, as well as the length of worm (b_1), using Eqs. 21.1, 21.2, 21.3, 21.4 and 21.5, respectively (see Sect. 21.3).

The following example shows the details of applying the method.

Table 21.1 The number of worm threads (z_1) according to the speed ratio (u)

Speed ratio (u)	$u < 5$	$5 - 10$	$10 - 15$	$15 - 30$	$\mu \geq 30$
Number of worm threads (z_1)	6	4	3	2	1

Table 21.2 The allowable contact stresses of worm and wormgears (Wittel et al. 2013)

Wormgear material	Hardness (HB)	Allowable contact stress (σ_H, MPa)*
G-CuSn12	80	265
GZ-CuSn12	95	425
G-CuSn12Ni	90	310
GZ-CuSn12Ni	100	520
G-CuSn10Zn	75	350
GZ-CuSn10Zn	85	430
G-CuZn25Al5	180	500
GZ-CuZn25Al5	190	550
GZ-CuAl10Ni**	160	660
GJL-250**	250	350
GJS-400**	260	490

*For worms: if the worm is made of St, case-hardened and ground, σ_H is equal to that of the wormgear; if the worm is made of St, tempered and unpolished, σ_H is equal to 0.75 times of that of the wormgear; if the worm is made of St70 (E360), σ_H is equal to 0.6 times of that of the wormgear; if the worm is made of GJL, σ_H is equal to 0.5 times of that of the wormgear.
**For $v < 5$ m/s

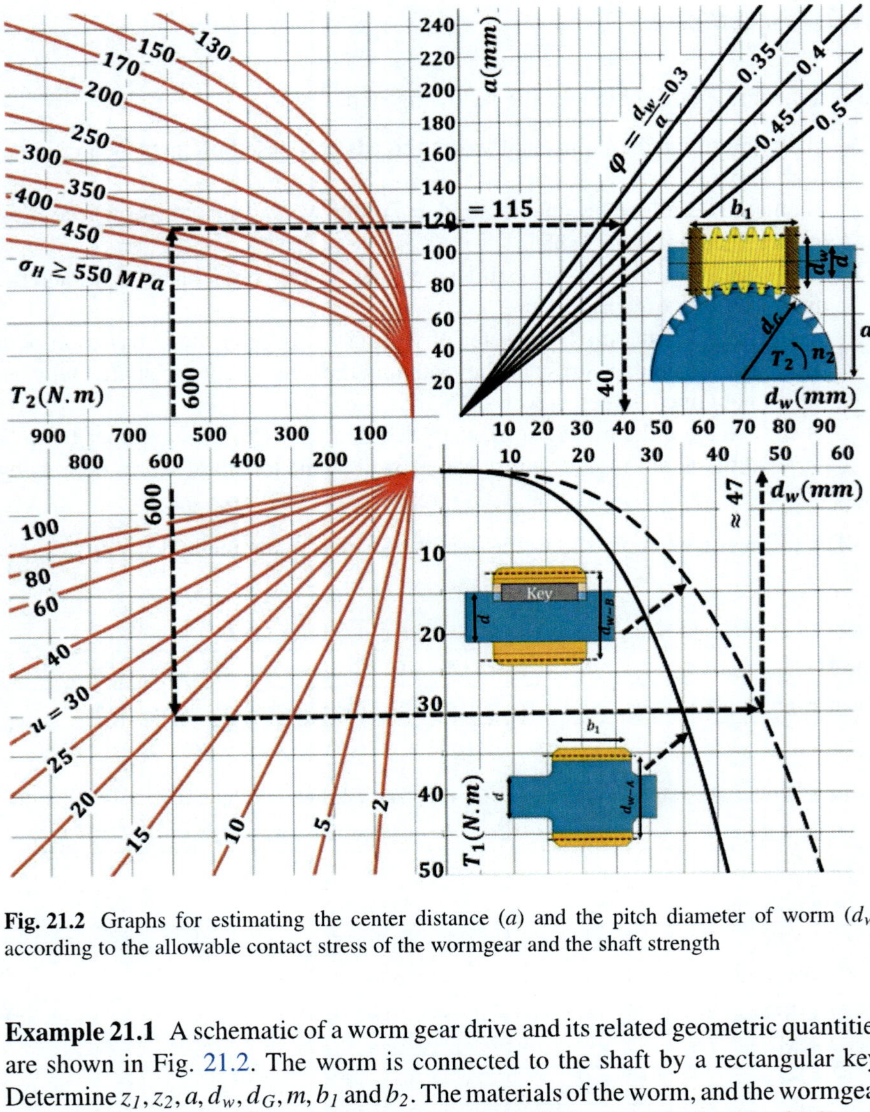

Fig. 21.2 Graphs for estimating the center distance (a) and the pitch diameter of worm (d_w) according to the allowable contact stress of the wormgear and the shaft strength

Example 21.1 A schematic of a worm gear drive and its related geometric quantities are shown in Fig. 21.2. The worm is connected to the shaft by a rectangular key. Determine $z_1, z_2, a, d_w, d_G, m, b_1$ and b_2. The materials of the worm, and the wormgear are tempered and unpolished steel, and GZ-CuSn12Ni, respectively (Fig. 21.3).

Solution

1. First, we determine the number of worm threads (z_1) from Table 21.1. Since the speed ratio is $u = 20$, the number of worm threads is:

$$z_1 = 2$$

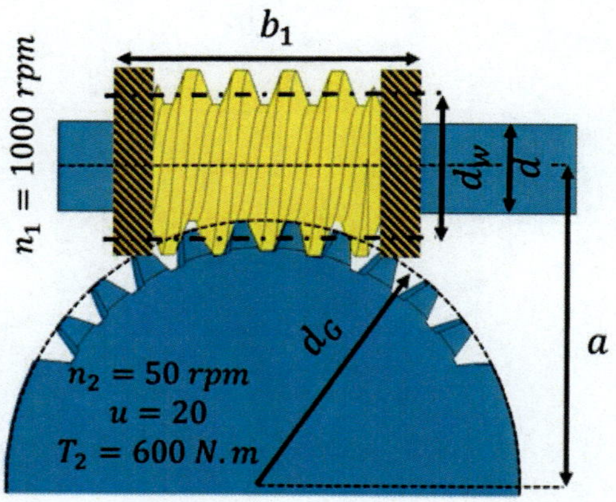

Fig. 21.3 an illustration for the wormgear drive of Example 21.1

2. Then, we should determine the pitch diameter of worm (d_w) and the center distance (a) from Fig. 21.2. For this purpose, we need the allowable contact stress of the teeth.

   We can find the allowable contact stress from Table 21.2.

   The allowable contact stress for the wormgear, which is made of GZ-CuSn12Ni is 520 MPa. Since the material of the worm is tempered and unpolished steel, its allowable contact stress based on Table 21.2 is:

$$\sigma_H = 0.75 \times 520 = 390 \text{ MPa}$$

   We use this value as the allowable contact stress for the estimations.

(a) Estimating the pitch diameter of the worm and the center distance based on the contact strength:

3. From the left horizontal axis of Fig. 21.2, we select the torque of $T_2 = 600$ N.m and go up to intersect the approximate curve associated with $\sigma_H = 390$ MPa.
4. Then, we go right to reach the upper vertical axis. The estimated value of the center distance can be found on the axis:

$$a = 115 \text{ mm}$$

5. Then, we continue going right until we cross the curve of $\varphi = 0.35$.

   φ is the ratio of the pitch diameter of worm to the center distance (Eq. 21.6 (see Sect. 21.3)). The designer selects the value of φ. Here, we arbitrarily select $\varphi = 0.35$.
6. Then, we go down to find the pitch diameter of the worm:

$$d_w = 40 \text{ mm}$$

(b) Determining the pitch diameter of the worm (d_w) based on the shaft strength:

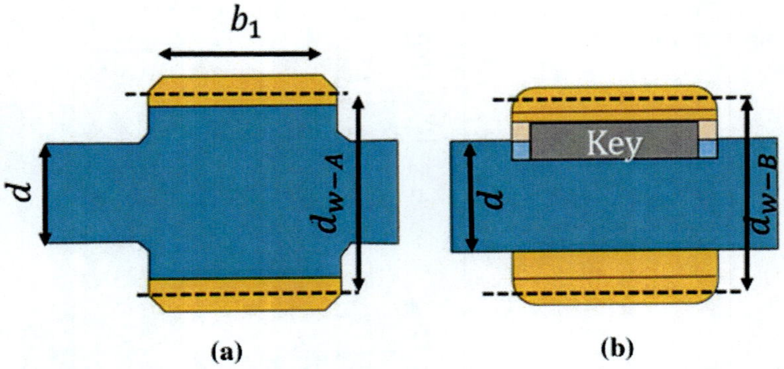

Fig. 21.4 The connection of the worm and shaft in wormgear drives, **a** the one-piece design, **b** the connection by keys

7. For this purpose, we should use Fig. 21.2.

   Note: if the worm is connected to the shaft by means of keys (like Fig. 21.4b), the condition $d_w \geq 2d$ should be fulfilled. However, in cases that the worm and shaft are one-piece (like Fig. 21.4a), the condition $d_w \geq 1.5d$ applies (Decker and Kabus 2014).

   Since in this example, the worm is connected to the shaft by a rectangular key, the pitch diameter of worm should be greater than or equal to the shaft diameter, $d_w \geq 2d$. This means that we should use the curve related to "d_{w-B}" in Fig. 21.2.

8. Form the left horizontal axis of Fig. 21.2, we select $T_2 = 600$ N.m, and go down to reach the curve of $u = 20$.
9. Then, we go right to cross the curve of "d_{w-B}" (the dashed line).
10. Then, going up to reach the right horizontal axis gives us the value of d_w:

$$d_w = 47\,\text{mm}$$

11. In the end, between the two obtained values for the pitch diameter of the worm, $d_w = 40$ mm and $d_w = 47$ mm, we select the higher one. Furthermore, for safety and increasing the rigidity, we choose a bigger size for the diameter, thus:

$$d_w = 50\,\text{mm}$$

12. Now, we can calculate the rest of the dimensions using Eqs. 21.1, 21.2, 21.3, 21.4 and 21.5, respectively (see Sect. 21.3), as follows:

    The number of wormgear teeth:

$$z_2 = uz_1 = 20 \times 2 = 40$$

The module of the wormgear (we should round the value of the module to a standard size using Table 21.3):

$$m = \frac{2a - d_w}{z_2} = \frac{2 \times 115 - 50}{40} = 4.5 \xrightarrow{\text{round to a standaed value}} m = 4.5 \text{ mm}$$

The diameter of the wormgear:

$$d_G = mz_2 = 4.5 \times 40 = 180 \text{ mm}$$

The length of worm:

$$b_1 = 2m\sqrt{z_2 + 1} = 2 \times 4.5 \times \sqrt{40 + 1} = 57.62 \text{ mm}$$

The wormgear width:

$$b_2 = 0.45d_w + 2.7m = 0.45 \times 50 + 2.7 \times 4.5 = 34.65 \text{ mm}$$

13. After rounding the module to a standard size, we have to recalculate the center distance (a). Therefore, according to Eq. 21.7 (see Sect. 21.3):

$$a = 0.5(d_G + d_w) = 0.5(180 + 50) = 115 \text{ mm}$$

Therefore the final results are:

$z_1 = 2$, $z_2 = 40$, $a = 115$ mm, $d_w = 50$ mm, $d_G = 180$ mm, $m = 4.5$ mm, $b_1 = 57.62$ mm and $b_2 = 34.65$ mm.

21.2.2 Method II (Suggested Method)

The steps of this method are listed below:

1. Determine the number of worm threads (z_1) from Table 21.1.
2. Calculate the number of wormgear teeth (z_2) using Eq. 21.3 (see Sect. 21.3) ($z_2 = uz_1$).
3. According to the material of worm and wormgear, specify the value of the coefficient K_s from Table 21.4.
4. Estimate an initial value for the pitch diameter of the wormgear (d_G) using Fig. 21.5.
5. Determine the pitch diameter of the worm (d_w), according to the shaft strength, from Fig. 21.2.

Table 21.3 Standard sizes for the module in wormgear drives (all values are in mm) Decker and Kabus; Maschinenelemente, 2014 Carl Hanser Verlag München)

Series 1	0.05	0.06	0.08	0.1	0.12	0.16	0.2	0.25	0.3	0.4	0.5	0.6	0.7	0.8	0.9	1	1.25
	1.5	2	2.5	3	4	5	6	8	10	12	16	20	25	32	40	50	60
Series 2	0.055	0.07	0.09	0.11	0.14	0.18	0.22	0.28	0.35	0.45	0.55	0.65	0.75	0.85	0.95	1.125	1.375
	1.175	2.25	2.75	3.5	4.5	5.5	7	9	11	14	18	22	28	36	45	55	70

Table 21.4 The material factor K_s, for worm gear drives (Böge 2011)

Worm material	Wormgear material	K_s
Steel, hardened and ground, (like E355, E360, C15, 16MnCr5)	CuSn alloys (like CuSn12Ni2-C)	8
	Al alloys (like CuZn25Al5Mn4Fe3-C)	4
	Perlite casting	12
Steel, tempered and not ground (like E355, E360, 42CrMo4)	CuSn alloys (like CuSn12Ni2-C)	5
	Al alloys (like CuZn25Al5Mn4Fe3-C)	2.5
	Zn alloys (like CuZn25Al5Mn4Fe3-C)	2
	Cast iron (like EN-GJL-150)	4
Cast iron (like EN-GJL-200)	CuSn alloys (like CuSn12Ni2-C)	4
	Al alloys (like CuZn25Al5Mn4Fe3-C)	2
	Cast iron (like EN-GJL-150)	3.5

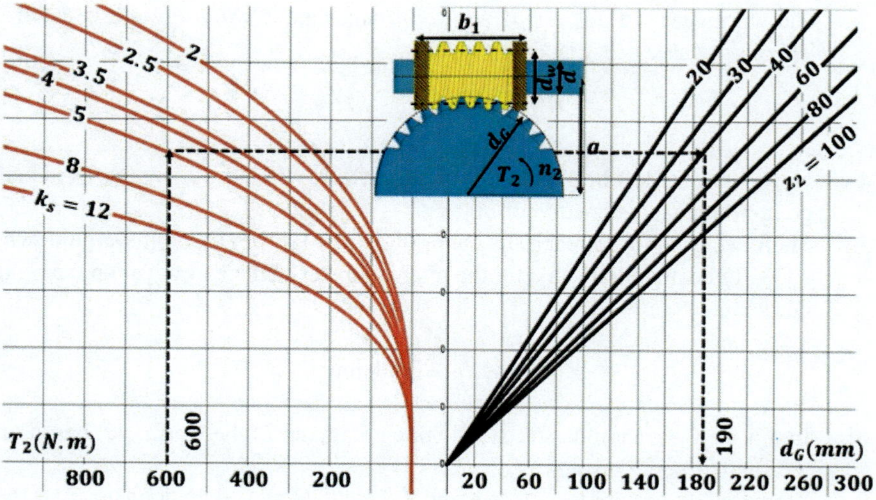

Fig. 21.5 Graphs for determining the pitch diameter of wormgears (dG)

6. Calculate the module (m) using Eq. 21.2 (see Sect. 21.3) $\left(m = \frac{d_G}{z_2}\right)$. Then, round it to a standard value from Table 21.3.
7. Calculate the final value of the wormgear pitch diameter (d_G) using the standard module and Eq. 21.2 (see Sect. 21.3) ($d_G = mz_2$).
8. Obtain the center distance (a) according to Eq. 21.7 (see Sect. 21.3) ($a = 0.5(d_G + d_w)$).
9. Calculate the wormgear width (b_2) and the length of worm (b_1) using Eqs. 21.4 and 21.5, respectively (see Sect. 21.3).

The following example describes how the method should be applied.

Example 21.2 Solve Example 21.1 using Method II.

Solution

1. First, we should specify the number of worm threads, z_1, from Table 21.1. As the speed ratio is $u = 20$, the number of threads should be:

$$z_1 = 2$$

2. Then, we calculate the number of wormgear teeth according to Eq. 21.3 (see Sect. 21.3):

$$z_2 = uz_1 = 20 \times 2 = 40$$

3. Then, we find the coefficient K_s from Table 21.4. The material of the worm and wormgear is *Steel, tempered (not ground)* and *CuSn alloys*, respectively. Therefore value of K_s is:

$$K_S = 5$$

4. Now, from the left horizontal axis of Fig. 21.5, we select $T_2 = 600$ N.m and go up to intersect the curve of $K_s = 5$.
5. Then, we go right to reach the curve of $z_2 = 40$. From there, if we go down, we will find an initial value for the pitch diameter of the wormgear on the right horizontal axis:

$$d_G = 190\,\text{mm}$$

6. In this step, we should estimate the pitch diameter of the worm, d_w, based on the shaft strength, using Fig. 21.2.
7. From the left horizontal axis, we find $T_2 = 600$ N.m, and go down to cross the curve of $u = 20$.
8. Then, we go right to reach the curve related to "d_{w-B}" (the dashed line).

   As the worm is connected to the shaft by a rectangular key (like Fig. 21.4b), we should use the curve of "d_{w-B}".

9. Then, we go up to find the approximate value of d_w:

$$d_w = 47 \text{ mm}$$

For safety and increasing the rigidity, we choose of value of $d_w = 50$ mm for the diameter.

10. Now, we can calculate the module from Eq. 21.2 (see Sect. 21.3), and round it up to a standard size using Table 21.3:

$$m = \frac{d_G}{z_2} = \frac{190}{40} = 4.75 \xrightarrow{\text{round to a standaed value}} m = 5\text{mm}$$

11. We should update the size of the wormgear pitch diameter with the standard module ($m = 5$ mm), by applying Eq. 21.2 (see Sect. 21.3):

$$d_G = mz_2 = 5 \times 40 = 200 \text{ mm}$$

12. The value of the center distance (a) can be obtained from Eq. 21.7 (see Sect. 21.3):

$$a = 0.5(\mathrm{d_G} + \mathrm{d_w}) = 0.5 \times (50 + 200) = 125 \text{ mm}$$

13. The wormgear width (b_2) and the length of worm (b_1) can be calculated using Eqs. 21.4 and 21.5, respectively (see Sect. 21.3):

$$b_2 = 0.45d_w + 2.7m = 0.45 \times 50 + 2.7 \times 5 = 36 \text{ mm}$$
$$b_1 = 2m\sqrt{z_2 + 1} = 2 \times 5 \times \sqrt{z_2 + 1} = 64 \text{ mm}$$

21.3 Equations

$$m = \frac{2a - d_w}{z_2} \quad (21.1)$$

$$m = \frac{d_G}{z_2} \quad (21.2)$$

$$u = \frac{n_1}{n_2} = \frac{T_2}{T_1} = \frac{z_2}{z_1} \quad (21.3)$$

$$b_2 = 0.45\, d_w + 2.7\, m \quad (21.4)$$

$$b_1 = 2m\sqrt{z_2 + 1} \quad (21.5)$$

$$\varphi = \frac{d_w}{a} \tag{21.6}$$

$$a = 0.5(d_G + d_w) \tag{21.7}$$

The equations of Fig. 21.2:

$$\begin{aligned} a &= 750\sqrt[3]{\frac{T_2}{\sigma_H^2}} \\ d_{w-A} &\approx 1.5 \times 7.5^3\sqrt{\frac{T_2}{u}} \\ d_{w-B} &\approx 2 \times 7.5^3\sqrt{\frac{T_2}{u}} \end{aligned} \tag{21.8}$$

The equations of Fig. 21.5:

$$d_G = 11^3\sqrt{\frac{T_2 z_2}{K_s}} \tag{21.9}$$

The equations of Zhukov and Gurevich (2014):

$$\begin{aligned} a &\approx 645^3\sqrt{\frac{T_2}{\sigma_H^2}} \\ d_{G-max} &= 1000^3\sqrt{\frac{T_2}{\sigma_H^2}} \\ m_{max} &= \frac{2a}{1.25z_2} \\ d_{w-min} &= 0.4a \\ d_{w-max} &= 8.3^3\sqrt{T_2} \end{aligned} \tag{21.10}$$

References

Böge A (2011) Handbuch Maschinenbau: Grundlagen und Anwendungen der Maschinenbau-Technik. Vieweg+Teubner, Wiesbaden (in German)

Decker K-H, Kabus K (2014) Decker Maschinenelemente Funktion. Carl Hanser Verlag, München (in German), Gestaltung und Berechnung

Jiang W (2019) Analysis and design of machine elements, 1st edn. Wiley, New Jersey

Mott RL, Vavrek EM, Wang J (2018) Machine elements in mechanical design, 6th edn. Pearson Education, New York

Ugural AC (2015) Mechanical design of machine components, 2nd edn. CRC Press, Boca Raton
Wittel H, Muhs D, Jannasch D, Voßiek J (2013) Roloff/Matek Maschinenelemente Normung, Berechnung, Gestaltung. Springer Vieweg, Wiesbaden (in German)
Zhukov KP, Gurevich YE (2014) Design of parts and assemblies of machines: a textbook for universities. Mechanical Engineering, Moscow (in Russian)

Chapter 22
Rolling Contact Bearings

Nomenclature

b	Width of bearing, mm
C	Basic dynamic load rating, kN or N
C_0	Basic static load rating, kN or N
d	Shaft diameter, mm
D_b	Outer diameter of bearing, mm
d_b	Inner diameter of bearing, mm
F	Equivalent radial load from attached elements to the shaft, N
F_a	Axial force, kN or N
F_t	Tangential force, kN or N
F_r	Radial force, kN or N
H	Bearing life in hours, h
m	A coefficient for rough estimation of shaft diameter
n	Rotational speed of shaft, rpm
P	Equivalent load on bearing, kN
T	Torque, N.m
u	Speed ratio
β	Helix angle, degree
δ	Pitch cone angle of bevel gears, degree

M. Yaghoubi and H. Tavakoli, *Mechanical Design of Machine Elements by Graphical Methods*, Materials Forming, Machining and Tribology,
https://doi.org/10.1007/978-3-031-04329-1_22

22.1 Introduction

Rolling contact bearings (also known as rolling bearings) are machine elements that are most commonly designed to support rotating shafts. The essential parts of a rolling bearing are an outer ring (race), an inner ring (race), rolling elements (balls or rollers), and a separator (or retainer or cage). The outer ring is usually stationary and is fixed to the machine body. The inner ring is pressed onto the shaft and rotates with it. The rolling elements roll between the outer and inner rings, and permit a low-friction rotation of the shaft. The retainer spaces the rolling elements and prevents them to rub each other during operation (Juvinall and Marshek 2011; Klebanov and Groper 2016; Mott et al. 2018; Jiang 2019).

Rolling bearings have advantages such as high efficiency and reliability, minimum maintenance, low starting friction, ability to support radial, thrust, or combined radial and thrust loads, etc. (Collins et al. 2009).

One of the most important characteristics of rolling bearings to be specified is their bearing life. To determine the variable, the forces applied to the bearings should be calculated. This needs a good knowledge of *statics* and referring to manufacturers' catalogs, which is a time-consuming process. Instead, in this chapter we propose graphical methods for easy estimation of the bearing life and subsequently the tolerance.

22.2 Design Methods

Cyclic contact stresses occur in rolling bearings during their operation. Accordingly, their life is limited because a fatigue failure caused by the stresses is inevitable. Therefore, the term bearing life is defined for rolling bearings as the number of revolutions or hours that they can operate at a uniform speed before the fatigue failure occurs (Ugural 2015).

The procedure of rolling bearings design is first to find the loads on them and calculate the equivalent load, then to determine the bearing life and specify the suitable bearing type. The following subsections provide details about our proposed graphical method for this purpose.

22.2.1 Equivalent Load

For design of a rolling bearing, it is necessary to find the axial (F_a) and radial (F_r) forces and consequently the equivalent force (P). Graphs presented in Fig. 22.1 can easily be applied for estimating the equivalent load of different rolling bearings.

The following example illustrates the method.

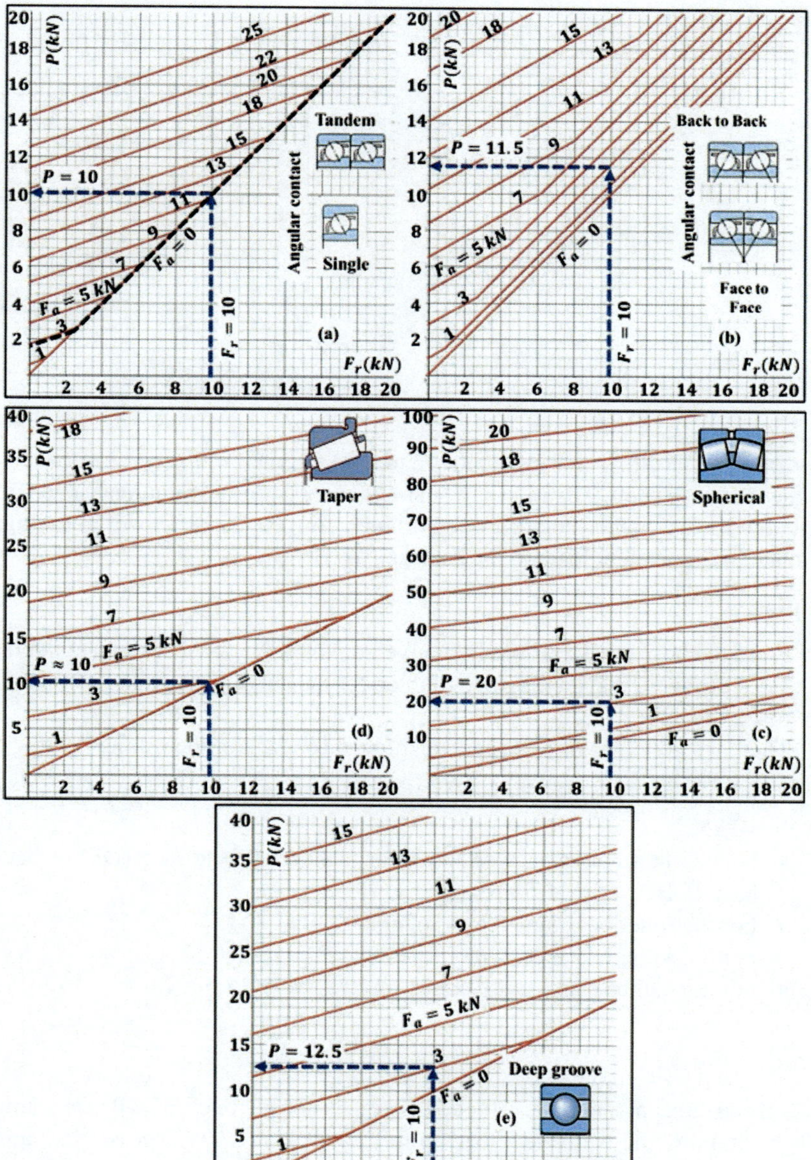

Fig. 22.1 Graphs for estimating the equivalent load on different rolling bearings, **a** ball bearings, type "Angular contact, tandem or single", **b** ball bearings, type "Angular contact, face to face or back to back", **c** roller bearings, type "Tapered", **d** roller bearings, type "Spherical", and **e** ball bearings, type "Deep groove"

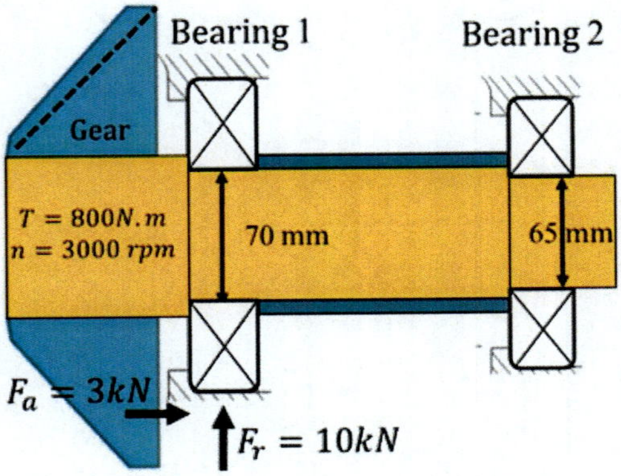

Fig. 22.2 An illustration for Example 22.1

Example 22.1 Rolling bearings are used to hold the shaft of Fig. 22.2. A bevel gear is attached to the shaft. The values of shaft diameters, torque, rotational speed of the shaft, and the axial and radial forces on the bearing No. 1 are shown in the figure. Estimate the equivalent load on the bearing No. 1.

Solution

1. From the horizontal axis of all the five graphs of Fig. 22.1, we select the radial force of $F_r = 10$ kN, and go up to intersect the curves related to the axial force of $F_a = 3$ kN.
2. Then, we go left to find the equivalent load for each of the rolling bearings:

   - For the ball bearings, type "Angular contact, tandem (DT) or single": $P =$ 10 kN
   - For the ball bearings, type "Angular contact, face to face (DF) or back to back (DB)": $P = 11.5$ kN
   - For the roller bearings, type "Spherical": $P = 20$ kN
   - For the roller bearings, type "Tapered": $P = 10$ kN
   - For the ball bearings, type "Deep groove": $P = 12.5$ kN

Notes:

- As it can be seen in Fig. 22.1, for all the graphs, the curve of $F_a = 0$ is in common.
- For obtaining the equivalent load from Fig. 22.1, it is not necessary to apply a safety factor. Because the graphs have been already created conservatively.

22.2.2 Bearing Life

Here, we specify the suitable rolling bearing for an application using Figs. 22.3 and 22.4 for roller and ball bearings, respectively. This is done based on the expected life of rolling bearings. Each rolling bearing has a specific life and when it is installed

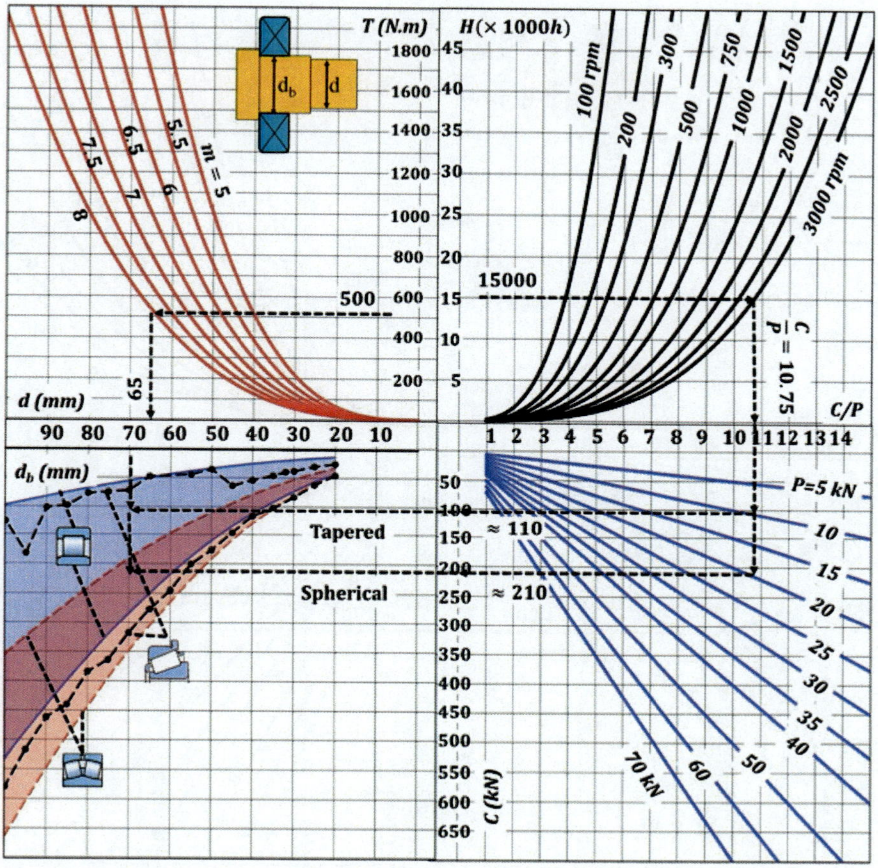

Fig. 22.3 Graphs for determining the type of roller bearings and the bearing life

on the machine, the life should be matched with the working hours of the machine. We explain the procedure within the following example.

Example 22.2 Assume that the life of the bearing No. 1 in Example 22.1, with expected work of 8 h per day, should be 5 years. Find the suitable bearing type for this application. Suppose that the torque on the shaft is $T = 500$ N.m.

Solution

1. First, we should calculate the bearing life in hours:

$$H = \text{year} \times 365 \times \text{hours per day} = 5 \times 365 \times 8 \approx 15000 \text{ hours}$$

2. Then, we should estimate the minimum shaft diameter (d) and the inner diameter of the bearing (d_b) from Figs. 22.3 and 22.4.

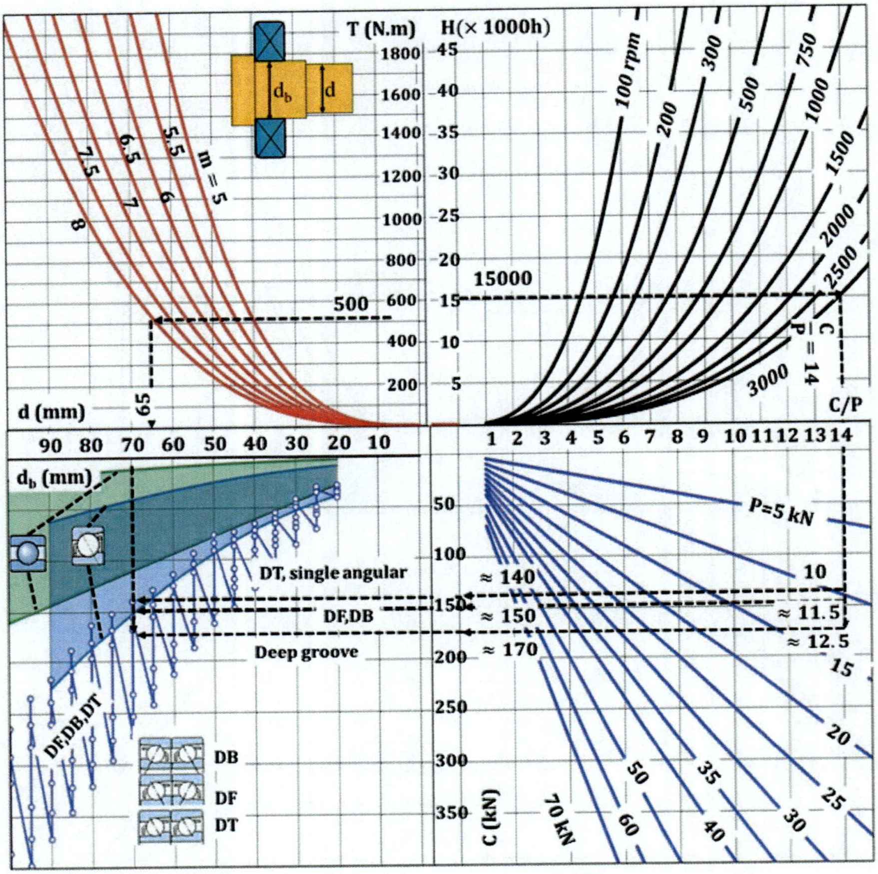

Fig. 22.4 Graphs for determining the type of ball bearings and the bearing life

3. In the second quadrants (the upper left corners) of Figs. 22.3 and 22.4 which are related to the roller and ball bearings, respectively, we select the torque of $T = 500$ N.m from the vertical axes, and go left to intersect the curves related to m = 8.

   According to Table 22.1, for the shaft of this example: $m = 8$.

Table 22.1 The coefficient m for rough estimation of the shaft diameter in Figs. 22.3 and 22.4

Shaft arrangement	Pinion shaft for bevel gears	Worm shaft for worm reducers	Cantilever shaft	Overhung or simply supported shaft	Pure torsion
m	8	7.5	6.5–7	5.5–6	5.5

4. Then, we go down to find the minimum shaft diameter:

$$d = 65\,\text{mm}$$

Due to the shoulder on the shaft for seating the bearing, we choose an inner diameter of $d_b = 70$ mm for the bearing.

5. Now, we should check what types of bearings with an inner diameter of $d_b =$ 70 mm are available.
6. In the first quadrants (the upper right corners) of Figs. 22.3 and 22.4, we select $H = 15{,}000$ h from the vertical axes, and go right to cross the curves related to $n = 3000$ rpm (as the rotational speed of the shaft in this example is 3000 rpm).
7. Then, we go down to determine the ratio of $\frac{C}{P} = 10.75$ for roller bearings (Fig. 22.3) and $\frac{C}{P} = 14$ for ball bearings (Fig. 22.4).
   (The ratio is used to determine the bearing tolerance.)
8. Then, we continue going down in both the figures to intersect the curves related to the equivalent forces obtained in Example 22.1, i.e. $P = 10$ kN and $P = 20$ kN in Fig. 22.3, and $P = 10$ kN, $P = 11.5$ kN and $P = 12.5$ kN in Fig. 22.4.
   For your convenience, we list the results of Example 22.1 below again:

- For the ball bearings, type "Angular contact, tandem (DT) or single": $P =$ 10 kN
- For the ball bearings, type "Angular contact, face to face (DF) or back to back (DB)": $P = 11.5$ kN
- For the roller bearings, type "Spherical": $P = 20$ kN
- For the roller bearings, type "Tapered": $P = 10$ kN
- For the ball bearings, type "Deep groove": $P = 12.5$ kN

9. Then, from all the intersected points, we go left to find the basic load ratings (C) for each of the bearing types on the vertical axes:

   In Fig. 22.3:

- For the roller bearings, type "Spherical": $C = 210$ kN
- For the roller bearings, type "Tapered": $C = 110$ kN

   In Fig. 22.4:

- For the ball bearings, type "Angular contact, tandem (DT) or single": $C =$ 140 kN
- For the ball bearings, type "Angular contact, face to face (DF) or back to back (DB)": $C = 150$ kN
- For the ball bearings, type "Deep groove": $C = 170$ kN

10. Therefore, the rolling bearing with an inner diameter of $d_b = 70$ mm, should have a basic load rating greater than the values above to be able to endure for 15,000 h.

As the next step, we should check which one of the bearing types with the inner diameter of $d_b = 70$ mm has this condition.

11. After finding the basic load ratings on the lower vertical axes of Figs. 22.3 and 22.4, we keep going left. Simultaneously, from the left horizontal axes, we select the diameter of $d_b = 70$ mm and go down.

    In Fig. 22.3:

    The intersection points for the *spherical* and *tapered* bearings are in their own range (the load capacity range) (the areas between the red dashed curves, and between the black dashed curves in the third quadrant of the figure are related to the ranges for spherical and tapered bearings, respectively). This means that if you refer to a manufacturer's catalog, you can find a spherical or tapered bearing with an inner diameter of $d_b = 70$ mm and a bearing life of 15000 hours or more. In addition, since the intersection points for both the spherical and tapered bearings are close to the lower limit of their range, the reliability of these bearings for this design is high.

    In Fig. 22.4:

    The intersection point for the *deep groove* ball bearing is out of its range (the area between the green solid curves in the third quadrant of the figure is related to the range for deep groove bearings). This means that this type of bearing is not suitable for the application of this example. In case of the *single angular* ball bearing, the intersection point is in the range, however, it is very close to the upper limit of the range (the area between the blue solid curves in the third quadrant of the figure is related to the range for single angular bearings). This indicates that the reliability of this type of bearing for the purpose of this example is very low. Nevertheless, the intersection points for the bearing types, *DT, DB*, and *DF* are in their range area and close to the lower limits of their range, which means that the reliability of these types of bearings are very high.

In conclusion, the suitable rolling bearings for the bearing No. 1 of Fig. 22.2 are spherical and tapered roller bearings, and DT, DB and DF ball bearings.

22.2.3 Allowable Shaft Speed for Rolling Bearings

When a rolling bearing is chosen, the allowable speed of shaft for the bearing should be taken into account. The shaft speed should not exceed that limit. The speed limit can be specified by referring to the bearing manufacturer's catalog. However, we propose diagrams in Fig. 22.5 for an easy check of the allowable speed. The following example shows the use of the diagrams.

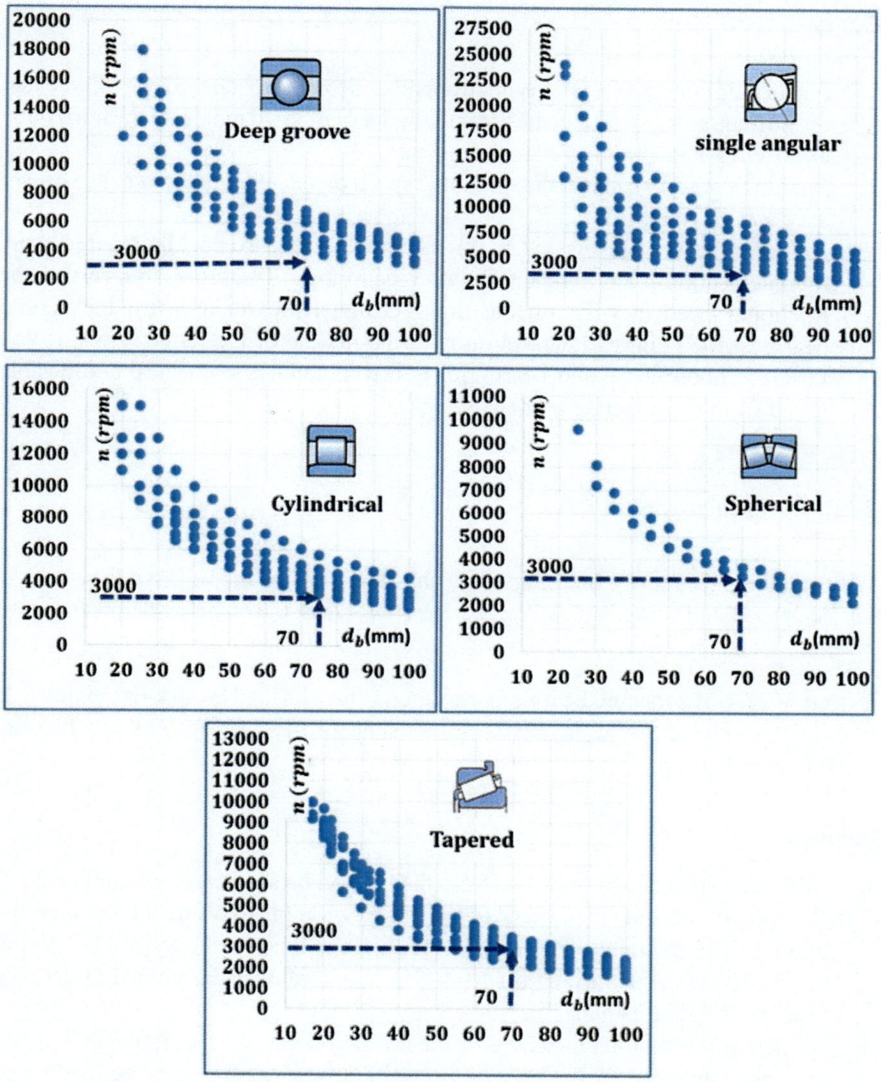

Fig. 22.5 Diagrams for checking the allowable speed of different bearing types (the assumption is that the bearings are lubricated with grease)

Example 22.3 In Example 22.1, the rotational speed of the shaft and the bearing inner diameter are $n = 3000$ rpm and $d_b = 70$ mm, respectively. Check for which bearing types the shaft speed is appropriate.

Solution

1. For all diagrams of Fig. 22.5 (which are related to different bearings, as indicated on the diagrams), we select the inner diameter of $d_b = 70$ mm from the horizontal axes, and go up.
2. Simultaneously, we select the rotational speed of $n = 3000$ rpm from the vertical axes and go right.
3. As it can be seen in Fig. 22.5, for all the bearings, except the tapered type, the intersection point is below the specified range. Therefore, in terms of the rotational speed, they are suitable for this design. However, in case of the tapered bearings, the point of intersection is in the middle of the range. Thus, if you want to choose a tapered bearing, it is recommended to check the allowable speed from the manufacturer's catalog.

22.2.4 *Tolerances*

With obtaining the ratio $\frac{C}{P}$ from Figs. 22.3 and 22.4, the tolerance class of the bearings can be determined using Tables 22.2, 22.3 and 22.4. See the example below for more details.

Example 22.4 If a tapered bearing is chosen for the bearing No. 1 of Example 22.1, find the tolerance class of the shaft and housing for the bearing. Assume that the housing is one-part.

Solution

1. We obtained the ratio $\frac{C}{P} = 10.75$ for the tapered bearing type in Example 22.2.
2. To specify the tolerance class of the shaft, we refer to Table 22.2. Since the ratio $\frac{\mathrm{C}}{\mathrm{P}}$ is in the range of $7 < \frac{C}{P} < 15$, and the inner diameter of the bearing ($d_b = 70$ mm) is in the range of 40–100 mm, according to the table, the tolerance class of the shaft is *m5*.
3. To determine the tolerance class of the housing bores, we use Table 22.3. Since the outer diameter of the bearing is constant, we can consider the loading of the outer ring as "point loading". In addition, as the bearing No.1 of Fig. 22.2 should be always fixed, we consider its assembly condition as "Outer ring not easily displaced". Thus, according to Table 22.3 the tolerance class for one-part housing is H6 (as shown in the table).

22.3 Supplementary Material

If you want to use a rolling bearing and its catalog is not available, measure its dimensions (the outer and inner diameters, and the width), and simply apply the

Table 22.2 The tolerance classes of shafts for different rolling bearings (modified from Autodesk Inc (2021))

Operating conditions	Examples of applications	Shaft diameter (mm)			Tolerance class
		Deep groove	Tapered and cylindrical	Spherical	
The inner ring under rotational loading (or undetermined loading direction)					
Small and variable load $\frac{C}{P} > 15$	Electrical instruments, fans	18–100	<40	–	j6
	Cutting machines, conveyors	100–200	40–140	–	k6
Medium and high load $7 < \frac{C}{P} < 15$	Common loading, cutting machines	<18	–	–	j5
	Turbines, electrical motors	18–140	<40	<40	k5
	Gear boxes, combustion engines	100–140	40–100	40–65	m5
	Pumps	140–200	100–140	65–100	m6
High load, shocks $\frac{C}{P} < 7$	Bearings for axles of rail trucks	–	50–140	50–100	n6
	Traction motors, rolling mills	–	140–500	100–500	p6
High mounting precision	Cutting machines	<18	–	–	h5
		18–100	<40	–	j5
		100–200	40–140	–	k5
Only axial load		All diameters			j6
The inner ring under static loading					
The inner ring must move easily over shaft		All diameters			g6
The inner ring does not have to move easily over shaft		All diameters			H6
Tapered bore bearing (class 0) (with adapter or withdrawal sleeve)					
Overall load		All shaft diameters			h9/IT5

relations presented in Supplementary Table 22.1 for a rough estimation of the basic load rating and the allowable speed of the bearing. Supplementary Example 22.1 shows how the relations can be used.

In addition, Supplementary Table 22.2 offers relations for calculating the axial and radial loads on bearings, and Supplementary Table 22.3 gives relations for estimating the equivalent force on the shaft's attached components.

Table 22.3 The tolerance classes of housing bores for rolling bearings (modified from Autodesk Inc (2021))

Operating conditions	Assembly conditions	Housing	Examples of applications	Tolerance class
Circumferential loading of the outer ring				
High shock loading $\frac{C}{F} < 7$	Outer ring is not sliding	One-part	Wheel hubs, connecting rod bearing	P7
Ordinary and high loading: $7 < \frac{C}{F} < 15$			Pulleys, bearing of crankshafts	N7
Small and variable load $\frac{C}{F} > 15$			Transport and tightening pulleys	M7
Undetermined way of loading (If the load direction on the outer ring is highly irregular or a rotating load and stationary load are applied alternatively)				
High shock load	Outer ring non-sliding	One-part	Traction motors	M7
High and ordinary load	Outer ring non-sliding from the right		Pump electrical motors, crankshafts	K7
Ordinary and small load	Outer ring sliding from the right		Pump electrical motors, fans	J7
Precision mountings				
Ordinary and small loading	Outer ring non-slid. from the right, outer ring sliding, outer ring easily sliding	One-part	Bearings for cutting machines	K6
			Ball bearing for grinding spindles	J6
			Small electric motors	H6
Point loading of the outer ring				
All load ratings	Outer ring is easily sliding	One or two-part	Common machinery, axle bearing	H8
		Two-part	Transmission shafting	H8
	Outer ring not easily displaced	One-part	High running accuracy required	H6 (J6)

(continued)

Table 22.3 (continued)

Operating conditions	Assembly conditions	Housing	Examples of applications	Tolerance class
	Outer ring not easily displaced, angular contact ball bearings and tapered roller bearings with adjusted outer ring	Two-part	Normal running accuracy	H7 (J7)
Heat input via shaft	Outer ring easily displaced	One or two-part	Great electrical motors with spherical roller bearings	G7

Table 22.4 The tolerance classes of shafts and housing bores for axial bearings (Autodesk Inc, 2021).

Bearing type	Loading type		Shaft diameter (mm)	Tolerance class
Tolerances of shafts for axial bearings				
Thrust ball	Only axial loading		All diameter	j6
Thrust spherical-roller	Only axial loading		All diameter	j6
	Axial and radial loading simultaneously	Point loading	All diameter	j6
		Circumferential loading or undetermined way of loading	<200	k6
			200–400	m6
			>400	N6
Tolerances of housing bores for axial bearings				
Thrust ball	Only axial loading, in ordinary mount, seat ring can be with radial clearance			H8
Thrust spherical-roller	Only axial loading, seat ring is mounted with radial clearance			H8
	Axial and radial loading simultaneously	Point loading or undetermined way of loading		H7
		Circumferential loading		M7

Supplementary Example 22.1 Assume that a deep groove ball bearing from Koyo company with dimensions $d_b = 70$ mm, $D = 110$ mm, and $b = 13$ mm is used on a machine. Estimate the allowable static and dynamic loads and the maximum allowable speed.

Supplementary Table 22.1 Relations for a rough estimation of the basic load rating and the allowable speed for rolling bearings (modified from Koyo Co.)

Bearing type	Basic load rating (in N) for $d_b \leq 100$mm		Allowable speed (rpm) (lubrication by grease)**
	Static*	Dynamic	
Deep groove ball bearing	$C_0 \geq 15 \times A$ $C_{0ave} \approx 17 \times A$	$C \geq 20 \times A$ $C_{ave} \approx 24 \times A$	$480000 < K < 580000$
Cylindrical roller bearing	$C_0 \geq 40 \times A$ $C_{0ave} \approx 52 \times A$	$C \geq 40 \times A$ $C_{ave} \approx 52 \times A$	$400000 < K < 600000$
Spherical roller bearing	$C_0 \geq 65 \times A$ $C_{0ave} \approx 72 \times A$	$C \geq 60 \times A$ $C_{ave} \approx 72 \times A$	$330000 < K < 500000$
Tapered roller bearing	$C_0 \geq 50 \times A$ $C_{0ave} \approx 67 \times A$	$C \geq 50 \times A$ $C_{ave} \approx 63 \times A$	$236000 < K < 300000$
Angular contact ball bearing	$C_0 \geq 20 \times A$ $C_{0ave} \approx 25 \times A$	$C \geq 35 \times A$ $C_{ave} \approx 37 \times A$	(depends on the type) $450000 < K < 850000$ or $600000 < K < 850000$ or $800000 < K < 850000$

*$A = 0.5(d_b + D)b$

**$K = 0.5(d_b + D)n$

C_{ave} = average of C

Solution

1. By referring to Supplementary Table 22.1, for the deep groove ball gearing, the following values are obtained:

$$\text{A} = 0.5(d_b + D)b = 0.5(70 + 110) \times 13 = 1170$$
$$\text{K} = 0.5(d_b + D)n = 90\text{n}$$

$$\text{The static load:} \begin{cases} \text{Minimum value}: C_0 \geq 15 \times A \rightarrow C_0 \geq 15 \times 1170 \\ \qquad \rightarrow C_0 \geq 17550\text{ N} \\ \text{Average value}: C_0 \approx 17 \times A \rightarrow C_0 \approx 17 \times 1170 \\ \qquad \rightarrow C_0 \approx 19890\text{ N} \end{cases}$$

Supplementary Table 22.2 Relations for calculating the axial and radial loads on rolling bearings

Rolling bearings arrangement

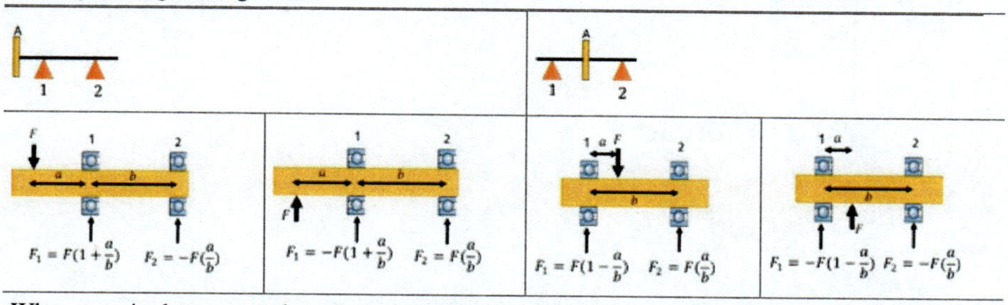

When two single-row angular contact ball bearings or two tapered roller bearings are used facing one another (see below), an axial component of force is produced under the rolling bearing No. 2. You can estimate the total axial load of the bearing No. 1 (F_{al}) as:

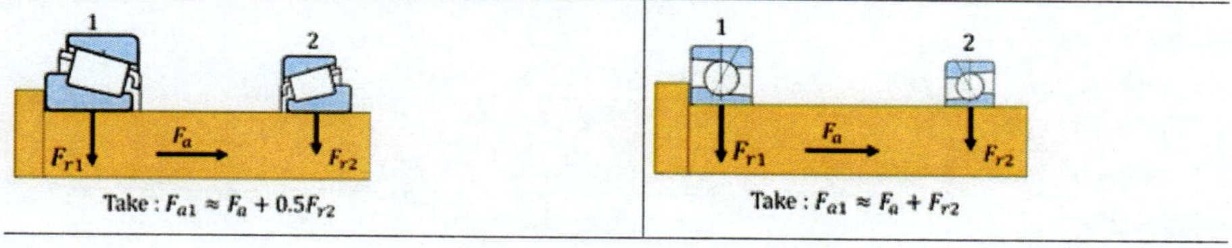

$$\text{The dynamic load:} \begin{cases} \text{Minimum value: } C \geq 20 \times A \rightarrow C \geq 20 \times 1170 \\ \qquad \rightarrow C \geq 23400 \text{ N} \\ \text{Average value: } C \approx 24 \times A \rightarrow C \approx 24 \times 1170 \\ \qquad \rightarrow C \approx 28080 \text{ N} \end{cases}$$

The allowable speed:

$$480000 < K < 580000 \xrightarrow{K=90n} 480000 < 90n < 580000 \rightarrow 5333\text{rpm} \leq n \leq 6444 \text{ rpm}$$

2. To be on the safe side, it is recommended to select a speed less than 5000 rpm.
3. For a bearing life of $H = 15{,}000$ hours (as in Example 22.2), the ratio $\frac{C}{P}$ must be greater than or equal to 14:

$$\frac{C}{P} \geq 14 \rightarrow P \leq \frac{C}{14} \xrightarrow{C=23400\text{N}} P \leq 1671 \text{ N or } 167 \text{ kg}$$

4. By using the results, we can check whether the bearing is appropriate for the work of the device or not.

Supplementary Table 22.3 Relations for estimating the equivalent force on the shaft's attached components

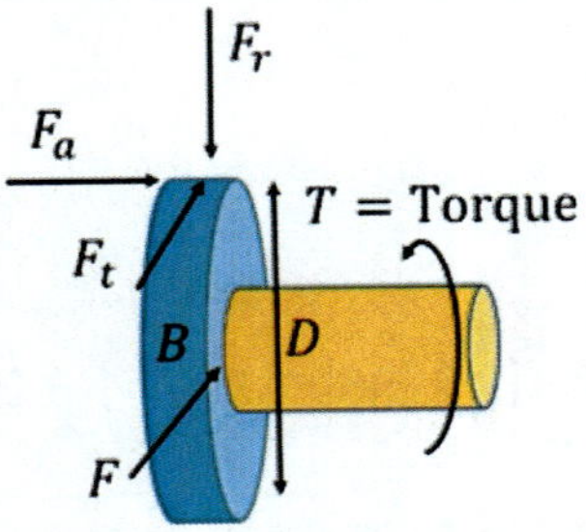

Component (B in the figure above)	Equivalent radial force (P) (all forces are in N)
Rigid coupling	$F = 0.2\ F_t$
Jaw coupling	$F \approx (18 - 43)\sqrt[3]{T}$
Flat belt	$\mathrm{F} = (2.5 - 4)F_t$
Timing belt and spur gear	$\mathrm{F} = (1.1 - 1.3)F_t$
V-belt	$\mathrm{F} = (1.5 - 2)F_t$
Chain	$\mathrm{F} = (1.2 - 1.5)F_t$
Output shaft of gearbox	$F = 200\sqrt{T}$
Input shaft of gearbox	$F = 150\sqrt{T}$
Bevel gear	$F_r = F_t\left(0.36\dfrac{\cos(\delta)}{\cos(\beta)} + \tan(\beta)\sin(\delta)\right)$ $F_a = F_t\left(0.36\dfrac{\sin(\delta)}{\cos(\beta)} - \tan(\beta)\cos(\delta)\right)$ $F = \sqrt{F_r^2 + F_t^2}$ $\delta = \tan^{-1}\left(\dfrac{1}{u}\right)$
$F_t = \frac{2000T}{D}$	

5. However, referring to the manufacturer's catalog gives the following values:

$$d_b = 70\ \text{mm}$$
$$D = 110\ \text{mm}$$
$$b = 13\ \text{mm}$$
$$C_0 = 21200\ \text{N}$$
$$C = 29700\ \text{N}$$
$$n < 6400\ \text{rpm}$$

22.4 Equations

The equations of Fig. 22.1.

For deep groove ball bearings:

$$P = \max(F_r, 0.56F_r + 2.3F_a)$$

For angular contact ball bearings, single or tandem:

$$P = \max(F_r, 0.35F_r + 0.57F_a)$$

For angular contact ball bearings, back to back or face to face:

$$P = \max(F_r + 0.55F_a, 0.57F_r + 0.93F_a)$$

For spherical roller bearings:

$$P = \max(F_r + 3F_a, 0.67F_r + 4.5F_a)$$

For tapered roller bearings:

$$P = \max(F_r, 0.4F_r + 2.1F_a) \tag{22.1}$$

References

Autodesk Inc. (2021) Engineer's Handbook. Retrieved from Autodesk Inventor support and learning center. https://knowledge.autodesk.com/

Collins J, Busby H, Staab G (2009) Mechanical design of machine elements and machines: a failure prevention perspective, 2nd edn. Wiley, New York

Jiang W (2019) Analysis and design of machine elements, 1st edn. Wiley, New Jersey

Juvinall R, Marshek K (2011) Fundamentals of machine component design, 5th edn. Wiley, New York

Klebanov BM, Groper M (2016) Power mechanisms for rotational and cyclic motions. CRC Press, Boca Raton

Mott RL, Vavrek EM, Wang J (2018) Machine elements in mechanical design, 6th edn. Pearson Education, New York

Ugural AC (2015) Mechanical design of machine components, 2nd edn. CRC Press, Boca Raton

Chapter 23
Hydraulic Systems

Nomenclature

D	Piston diameter, cm
d	Rod diameter, cm
d_i	Inner diameter of a pipe or a hose, mm
F	Hydraulic cylinder force, kg
H	Length of thread, mm
K	Hydraulic cylinder stroke factor
L_0	Base length of a hydraulic cylinder, mm
L_{close}	Closed length of a hydraulic cylinder, mm
L_{open}	Open length of a hydraulic cylinder, mm
P	Power, kW
p	Working pressure, bar
Q	Flow rate of a hydraulic pump, lit/min
S	Stroke of a hydraulic cylinder, mm
S_b	Base stroke of a hydraulic cylinder
t	Hydraulic cylinder thickness, mm
v	Hydraulic cylinder velocity, cm/s
z	Weld width, mm
σ_f	Fatigue strength, MPa
σ_y	Yield strength, MPa

M. Yaghoubi and H. Tavakoli, *Mechanical Design of Machine Elements by Graphical Methods*, Materials Forming, Machining and Tribology,
https://doi.org/10.1007/978-3-031-04329-1_23

23.1 Introduction

In some machines, hydraulic systems are used to transfer power and move some parts. In the hydraulic systems, a pump pressurizes a hydraulic fluid and sends it through hydraulic tubes to actuators (hydraulic motors and cylinders). There, the fluid's pressure is applied to move the parts or hold them in their position. Then, the fluid is moved through a filter, back to the pump, where it is pressurized again. The main advantage of hydraulic systems lies in their compactness and high efficiency.

In this chapter, we only discuss simple hydraulic systems that have hydraulic cylinders (Fig. 23.1).

23.2 Design Methods

To design hydraulic systems, the hydraulic cylinder's force, velocity and stroke, and the working pressure should be predetermined by the designer. Then, the piston diameter, the rod diameter, the pump flow rate, the pump pressure, and the hose diameter should be found.

In this chapter, we propose graphs for easy designing of hydraulic systems.

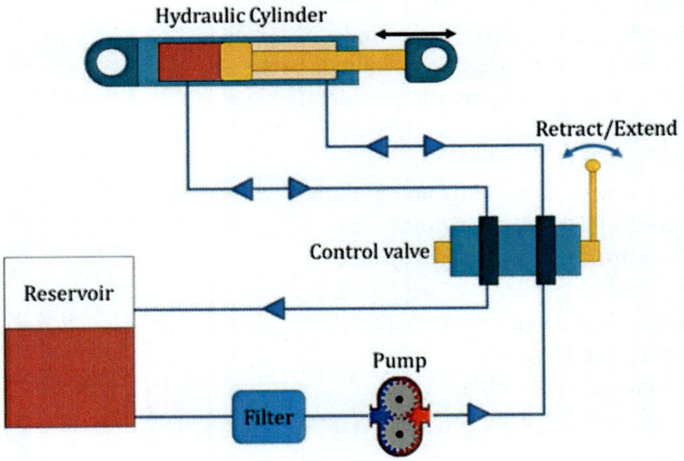

Fig. 23.1 A diagram of a simple hydraulic system

23.2.1 *The Piston and Rod Diameters of the Hydraulic Cylinder*

We assume that the cylinder applies force in both the extension and retraction strokes. Since the volume of the fluid, which is entering to the cylinder in the retraction stroke (part B in Fig. 23.2) is less than that in the extension stroke (part A in Fig. 23.2), we base the design on the cylinder retraction mode. Therefore, the ratio of the rod diameter (d) to the piston diameter (D) matters. An initial value for the ratio ($\frac{d}{D}$) should be chosen, that is the responsibility of the designer.

We propose a method for determining the piston and rod diameters using the graphs of Fig. 23.3, which are based on Eq. 23.1 (see Sect. 23.3).

The following example the use of the method.

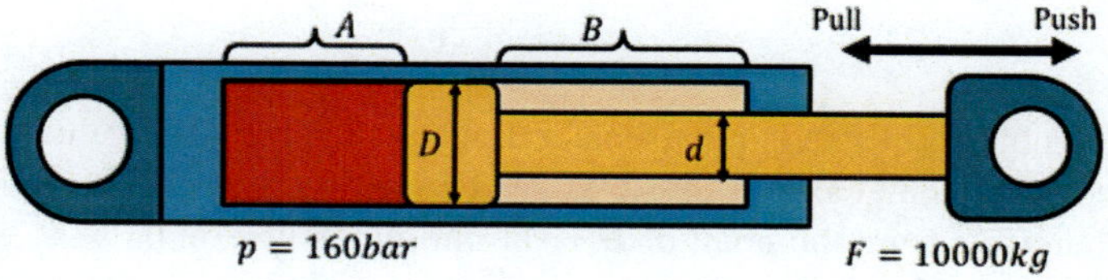

Fig. 23.2 A schematic of a hydraulic cylinder

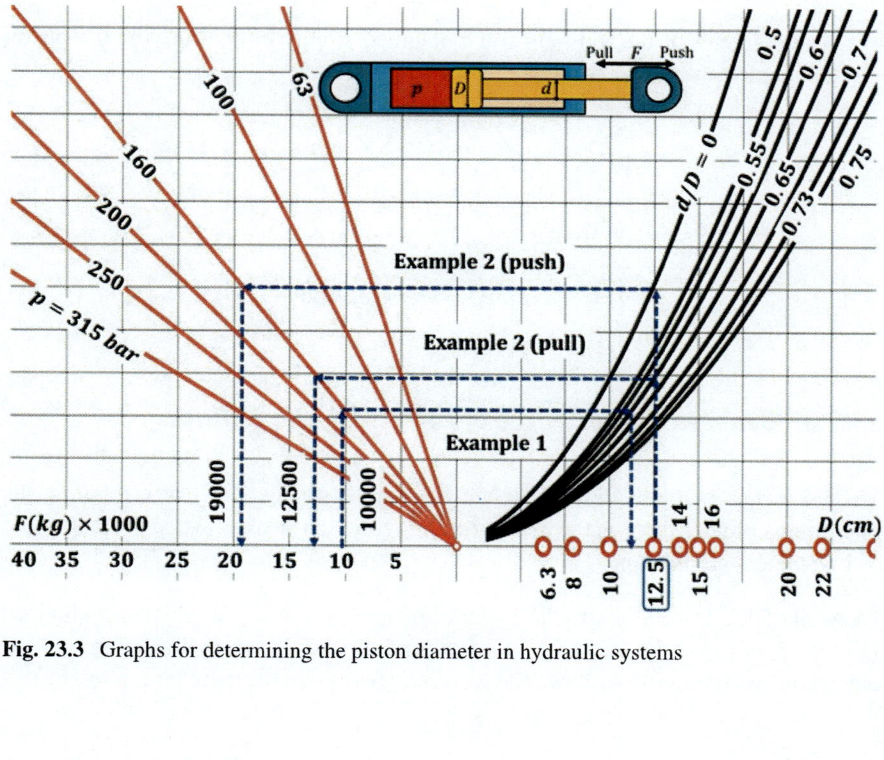

Fig. 23.3 Graphs for determining the piston diameter in hydraulic systems

Example 23.1 A hydraulic cylinder is schematically shown in Fig. 23.2. The cylinder applies a force of $F = 10{,}000$ kg in both the extension and retraction strokes. Determine the diameter of the piston (D) and the diameter of the rod (d). The working pressure is $p = 160$ bar.

Solution

1. In Fig. 23.3, we select the force of $F = 10{,}000$ kg from the horizontal axis and go up to intersect the curve of $p = 160$ bar.
2. Then, we go right to reach the curve related to $\frac{d}{D} = 0.6$ (we arbitrarily choose the ratio of $\frac{d}{D}$ equal to 0.6).
   Note: If the cylinder applies force in only one direction (extension stroke), the ration should be selected as $\frac{d}{D} = 0$, in Fig. 23.3.
3. Then, we should go down to find the piston diameter:

$$D = 11 \text{ cm}$$

The obtained value is not a standard piston diameter. The values shown on the right horizontal axis of Fig. 23.3 are the standard sizes.

Therefore, from the point of $D = 11$ cm on the horizontal axis, we should go right to reach the nearest standard diameter, which is:

$$D = 12.5 \text{ cm}$$

4. The diameter of the rod (d) can be obtained simply by multiplying the piston diameter by the ratio $\frac{d}{D}$:

$$d = \frac{d}{D} \times D \rightarrow d = 0.6 \times 12.5 = 7.5 \text{ cm or } 75 \text{ mm}$$

Therefore the results are:

$$D = 12.5 \text{ cm}, d = 7.5 \text{ cm}, \text{ and } \frac{d}{D} = 0.6$$

23.2.2 *The Maximum Force of the Hydraulic Cylinder*

The following example shows how we use graphs of Fig. 23.3 for estimating the maximum cylinder force in the extension and retraction strokes, when the piston and rod diameters are known.

Example 23.2 For the hydraulic cylinder of Example 23.1, in which we obtained $D = 12.5$ cm, $d = 7.5$ cm and $\frac{d}{D} = 0.6$, find the maximum cylinder force of the extension and retraction strokes. The working pressure in the cylinder is $p = 160$ bar.

Solution

1. From the right horizontal axis of Fig. 23.3, we select $D = 12.5$ cm and go up to cross both the graphs of $\frac{d}{D} = 0.6$ for the retraction stroke and $\frac{d}{D} = 0$ for the extension stroke.
2. Then, we go left to intersect the curve of the working pressure $p = 160$ bar.
3. Then, going down to the left horizontal axis gives us the maximum forces of the cylinder:
   In the retraction stroke: $F = 12500$ kg
   In the extension stroke: $F = 19000$ kg

Having these forces helps us to design other mechanical components of the hydraulic system such as the connection pins of cylinder head.

23.2.3 *The Hydraulic Pump Flow Rate and the Cylinder Speed*

Figure 23.4 presents graphs that can be used for determining the flow rate of hydraulic pumps, and the speed of the hydraulic cylinders. The graphs are based on Eq. 23.2 (see Sect. 23.3). For more details about the procedure of applying the graphs, refer to the examples below.

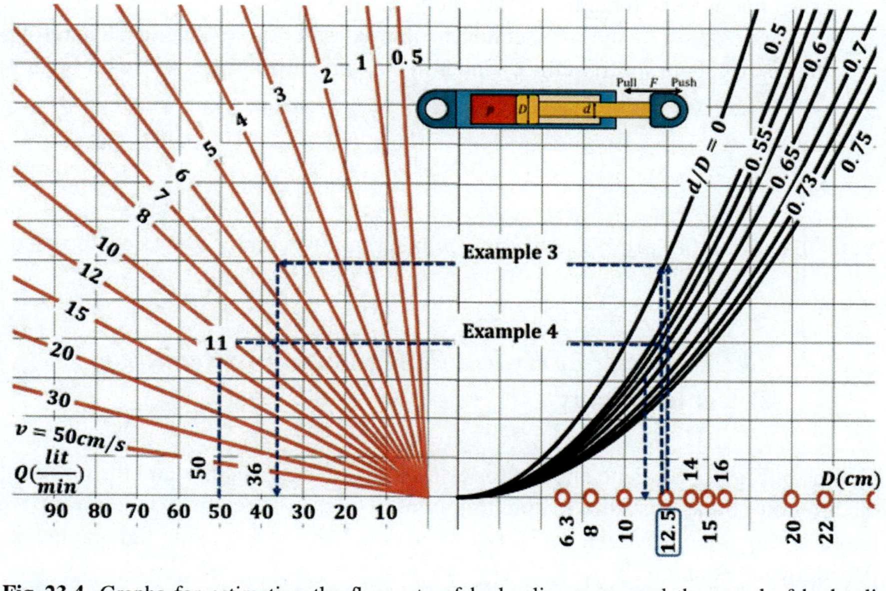

Fig. 23.4 Graphs for estimating the flow rate of hydraulic pumps and the speed of hydraulic cylinders

Example 23.3 Assume that for the hydraulic cylinder of Example 23.1, the designer has specified the speed of $v = 5$ cm/s, in the extension stroke. Determine the pump flow rate.

Solution

1. First, in Fig. 23.4, select $D = 12.5$ cm from the right horizontal axis, and go up to intersect the curve of $\frac{d}{D} = 0$, which is related to the cylinder extension stroke. Note that the designer has only specified the cylinder speed in the extension mode. Therefore, we have to do the estimation for this mode in which $\frac{d}{D} = 0$.
2. Then, we go left to reach the curve of $v = 5$ cm/s.
3. Then, going down to cross the left horizontal axis yields the minimum pump flow rate:

$$Q = 36 \text{ lit/min}$$

Example 23.4 Assume that in Example 23.3, a pump with a flow rate of $Q = 50$ lit/min is chosen. Determine the hydraulic cylinder speed in the retraction stroke.

Solution

1. In Fig. 23.4, from the right horizontal axis, we select $D = 12.5$ cm, and go up to cross the curve of $\frac{d}{D} = 0.6$, which is related to the cylinder retraction mode obtained in Example 23.1.
2. Then, we go left so that we cross the curves related to the cylinder speeds.
3. At the same time, we select $Q = 50$ lit/min from the left horizontal axis and go up so that we reach the line of the previous step.

   The location of the intersection is approximately $v = 11$ cm/s. Therefore, the approximate speed of the cylinder in the retraction stroke, with a pump flow rate of $Q = 50$ lit/min is:

$$v = 11 \text{ cm/s}$$

Note: to control the speed of actuators (cylinders), flow control valves can be used.

23.2.4 *The Diameter of Hydraulic Pipes and the Required Power of Pumps*

Graphs of Fig. 23.5 can be applied for estimating the size of pipes and the power of electromotors that drive the hydraulic pumps. The curves related to the power are based on Eq. 23.3 (see Sect. 23.3). For more details regarding using the graphs, see the following example.

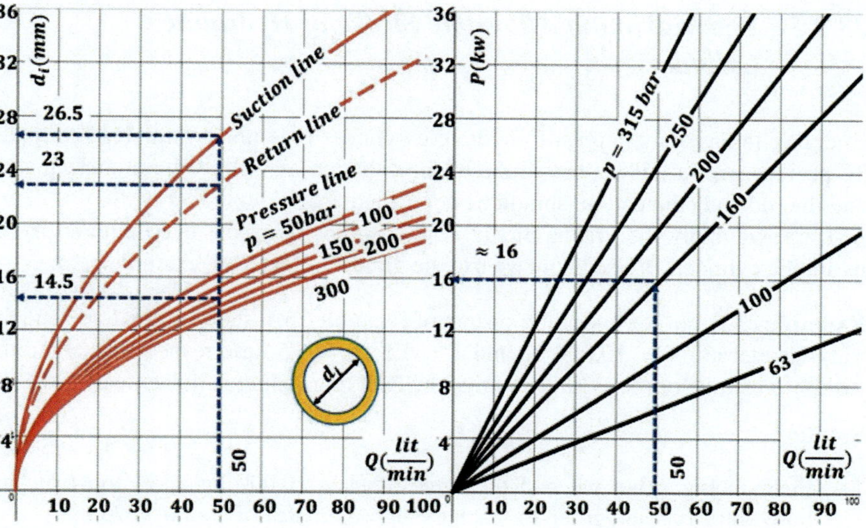

Fig. 23.5 Graphs for estimating the diameter of pipes or hoses and the power of electromotors that drive the hydraulic pumps

Example 23.5 With considering a pump flow rate of $Q = 50$ lit/min for the hydraulic system of Example 23.1, determine the power required for driving the pump (power of the electromotor) and the size of hydraulic pipes.

Solution

(a) Estimating the diameter of the pipes:

1. From the left horizontal axis of Fig. 23.5, we select $Q = 50$ lit/min and go up so that we cross the curves of $p = 160$ bar, "*Return line*" and "*Suction line*" (note that the working pressure is $p = 160$ bar).
2. Then, from all the three crossed points, we go left to find the inner diameter of the pipes or hoses as follows:

For $p = 160$ bar: $d_i \geq 14.5$ mm
For "*Return line*": $d_i \geq 23$ mm
For "*Suction line*": $d_i \geq 26.5$ mm

(b) Estimating the power of the electromotor:

3. From the right horizontal axis of Fig. 23.5, we find $Q = 50$ lit/min and go up to intersect the curve of $p = 160$ bar.
4. Then, we go left to find the minimum required power of the electromotor:

$$P = 16 \text{ kW}$$

23.2.5 *The Maximum Allowable Stroke of Hydraulic Cylinders*

Buckling failure may happen to hydraulic cylinder rods under compression loads. To prevent such a failure, the stroke of the cylinder, as well as the mounting style and the rod end connection, should be controlled.

Here, we apply the graphs of Fig. 23.6 for specifying the maximum stroke of hydraulic cylinders. To see how we use the graphs, refer to the example below.

Example 23.6 For the hydraulic system of Example 23.1, the working force and the rod diameter are $F = 10{,}000$ kg and $d = 7.5$ cm $= 75$ mm, respectively. Find the maximum stroke of the cylinder so that the buckling failure is prevented.

Solution

1. First, in Fig. 23.6, we find the force of $F = 10{,}000$ kg $= 10$ ton from the horizontal axis and go up to reach the approximate curve of $d = 75$ mm.
2. Then, we go right to obtain an initial value for the stroke:

$$S_b = 2500 \text{ mm}$$

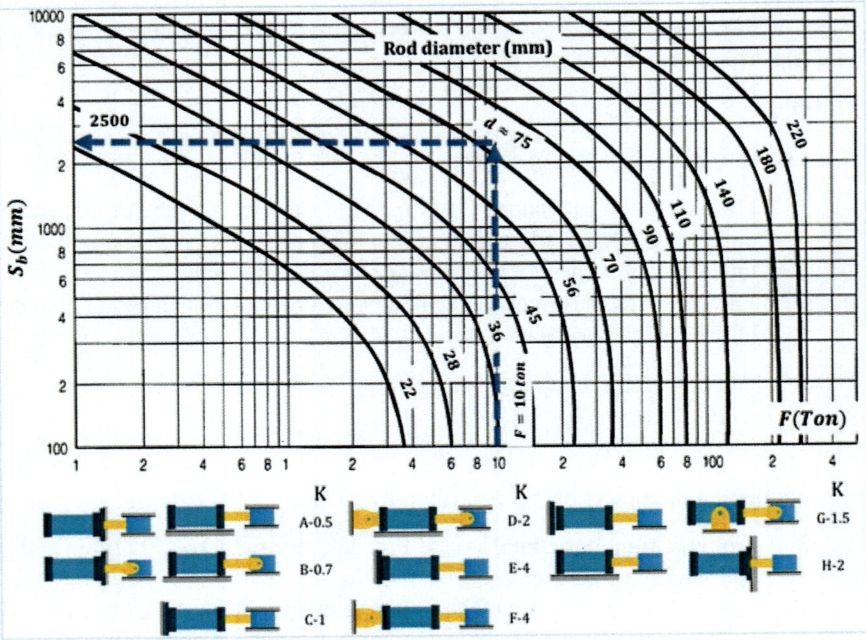

Fig. 23.6 Graphs for determining the maximum stroke of hydraulic cylinders (modified from Parker Hannifin Manufacturing Germany GmbH & Co. KG)

3. As can be seen in Fig. 23.2, the cylinder is mounted on the body and other element by means of pins on both ends (known as the eye mounting style). This is similar to the mounting style D in Fig. 23.6 with a stroke factor of $K = 2$.

   Therefore, for finding the real stroke (S), we should divide the obtained initial value ($S_b = 2500$ mm) by the stroke factor as follows:

$$S = \frac{S_b}{K} = \frac{2500}{2} = 1250 \text{ mm}$$

In conclusion, the stroke of the hydraulic cylinder should be less than or equal to 1250 mm ($S \leq 1250$ mm).

23.2.6 The Length of Hydraulic Cylinders

The graph of Fig. 23.8 can be employed for estimating the length of hydraulic cylinders that is also necessary for mounting them. The following example illustrate the method.

Example 23.7 The hydraulic cylinder of Example 23.1 with the piston diameter of $D = 12.5$ cm is supposed to be mounted between two press plates in the device (Fig. 23.7). Find the minimum distance of the plates (E) for mounting the cylinder. Consider the stroke as $S = 1000$ mm.

Solution

For correct mounting, the minimum distance between the plates (E) must be a little greater than the closed length of the cylinder (L_{close}). Therefore, we should find the value of L_{close}.

1. From the horizontal axis of Fig. 23.8, we select $D = 12.5$ cm and go up to intersect the curve.
2. Then, we go left to find the value of L_0:

$$L_0 = 450 \text{ mm}$$

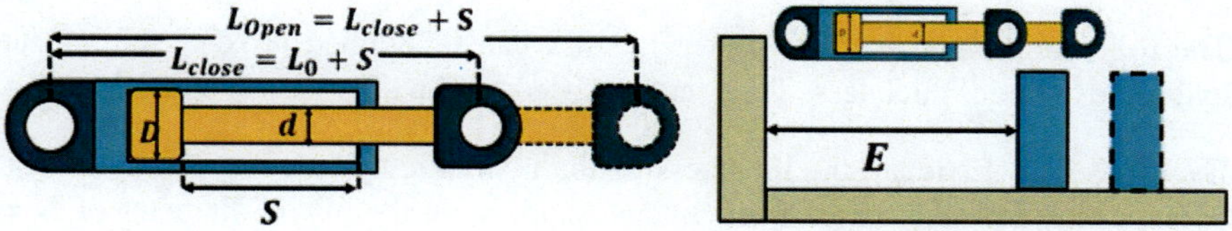

Fig. 23.7 An illustration for Example 23.7

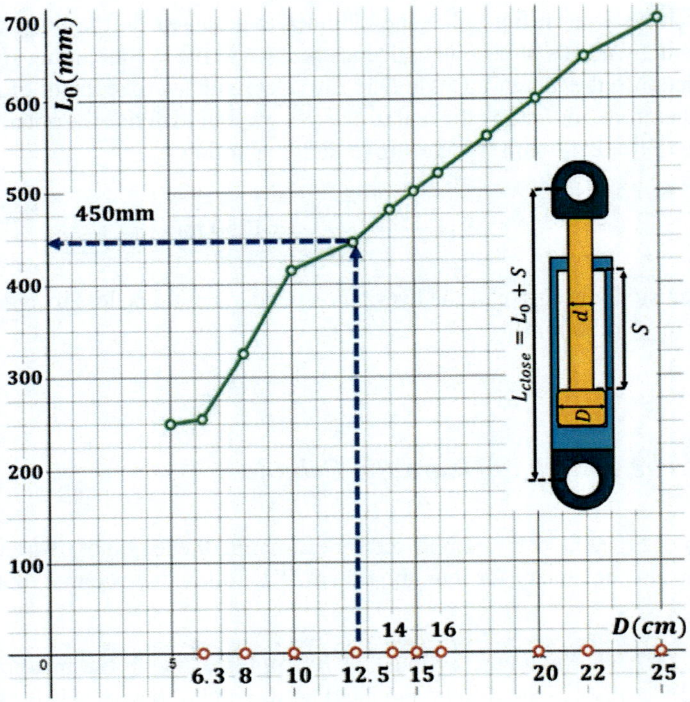

Fig. 23.8 A graph for estimating the length of hydraulic cylinders

3. Now the length of the cylinder in the extension and retraction, according to Fig. 23.7, is:

$$L_{close} = L_0 + S = 450 + 1000 = 1450 \text{ mm}$$
$$L_{open} = L_{close} + S = 1450 + 1000 = 2450 \text{ mm}$$

As a result, the distance E must be greater than the length of $L_{close} = 1450$ mm.

23.2.7 The Hydraulic Cylinder Thickness

The following example shows how Fig. 23.9 can be applied for determining the hydraulic cylinder thickness. The graph is based on Eq. 23.4 (see Sect. 23.3).

Example 23.8 Estimate the thickness of the hydraulic cylinder in Example 23.1. Assume that the material of the cylinder is st52 with the piston diameter of $D =$ 12.5 mm and the working pressure of $p = 160$ bar.

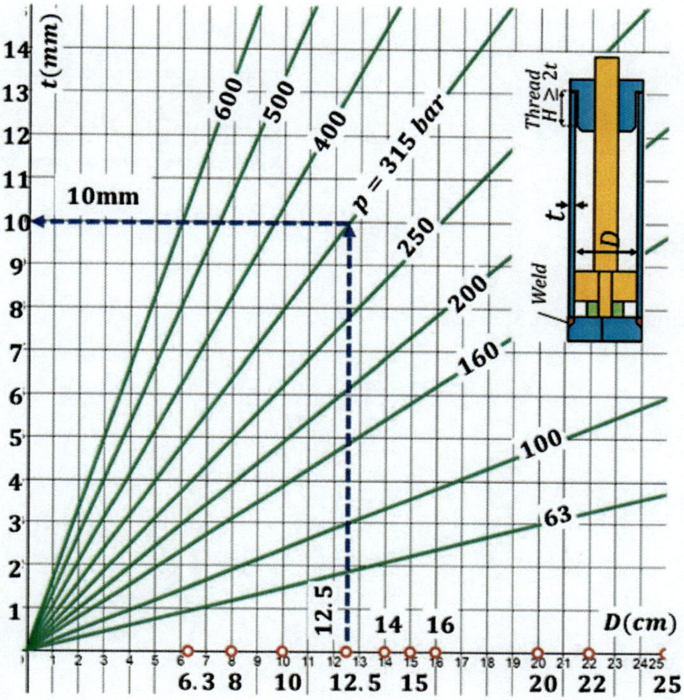

Fig. 23.9 A graph for estimating the thickness of hydraulic cylinder (the material is st52)

Solution

1. First, for safety, we multiply the working pressure by a factor of 2:

$$p = 2 \times 160 = 320 \text{ bar}$$

   Since we have a curve for the pressure of $p = 315$ bar in the figure, we select this pressure.
2. Using Fig. 23.9, we select $D = 12.5$ cm from the horizontal axis and go up to intersect the curve of $p = 315$ bar.
3. Then, we go left to find the minimum thickness of the cylinder tube (made of st52):

$$t = 10 \text{ mm}$$

Note: it is recommended that for determining the thickness of the cylinder, you select a pressure two times bigger than the working pressure; so that the weld and the thread in the ends of the cylinder (Fig. 23.10) are safe and are not needed to be calculated. However, some manufactures consider $t \approx 0.08D$ for hydraulic cylinders in working pressures up to $p = 250$ bar.

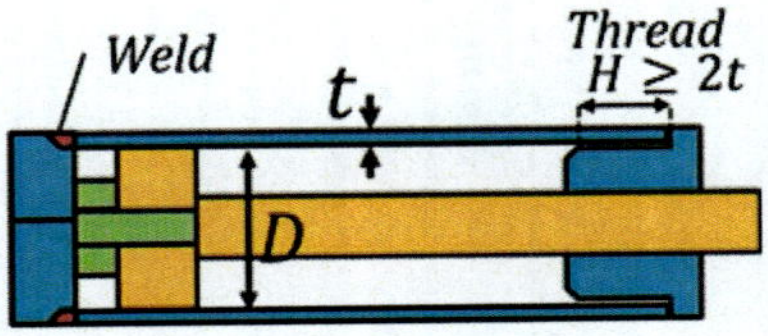

Fig. 23.10 A cross section view of a hydraulic cylinder with the weld and thread

23.3 Equations

The equation of Fig. 23.3:

$$F = \frac{\pi}{4}D^2\left(1 - \left(\frac{d}{D}\right)^2\right) \times p \tag{23.1}$$

The equation of Fig. 23.4:

$$Q = 0.06 \times \frac{\pi}{4}D^2\left(1 - \left(\frac{d}{D}\right)^2\right) \times v \tag{23.2}$$

The equation of Fig. 23.5:

$$P = \frac{pQ}{600 \times 0.85} \tag{23.3}$$

The equations of Fig. 23.9:

$$\begin{gathered} t = 10\frac{D}{2}\left(\sqrt{\frac{\sigma_f + 0.1p}{\sigma_f - 0.1p}} - 1\right) \\ \frac{10F}{\pi \times 0.8 \times 10 \times DH} \le 0.2\sigma_y \end{gathered} \tag{23.4}$$

Weld width of the cylinder and cap (if the cylinder material is st52):

$$z \ge \frac{D_p}{240}$$

Yield strength of st52: $\sigma_y = 355$ MPa.
Fatigue strength of st52 under pulsating load: $\sigma_f = 215$ MPa

Reference

Parker Hannifin Manufacturing Germany GmbH & Co. KG, Cylinder and Accumulator Division Europe, MMB Roundline Cylinders, Catalogue HY07–1215/UK (2006)